A Handy Guide to Paternity, Relationship, and Ancestry DNA Tests

Help in picking the right test

and in interpreting the results

D. Barry Starr, Ph.D.

Copyright © 2017 D. Barry Starr, Ph.D.

All rights reserved.

ISBN-10: 1544004982
ISBN-13: 978-1544004983

TABLE OF CONTENTS

	Page Number
Preface	4
DNA: Not So Mysterious	5
Part 1: DNA Basics	**6-36**
A Bit About DNA	7
DNA is Packaged in Chromosomes	13
Exceptions: Mitochondrial DNA, X, and Y	24
Shared DNA and Grandmas	27
About Genetic Tests	37
Part 2: Relationship Testing	**38-114**
DNA Relationship Test Advice	39
"Simple" Paternity Tests	43
Mutations Are More Common Than You Think	47
False Positives	49
My Brother is the Dad	51
Relationships other than Parent/Child	58
Paternity Based vs. Other Relationship Tests	60
Deeper Science Dive: From Four to One	67
Painting Your Chromosomes	72
Powerful Tests Can Find Fifth Cousins	76
Mom's 4th Cousin Shares No DNA with You	80

TABLE OF CONTENTS (cont.)

	Page Number
Teasing Extra Information Out of your Results	85
Are Your Parents Related?	86
Mom's DNA or Dad's?	93
X and the Y	95
Mitochondrial DNA	99
Half-Brothers and the Y	102
Relationship Tests Can Be Wrong (Chimerism)	107
Artificial Chimeras	113
Part 3: Ancestry: From Where Did I Come?	**114-177**
Some Ancestral Testing Advice	116
What Autosomal DNA Ancestry Tests Look For	117
Same Parents, Different Ancestry	128
55% Italian from One Parent	136
Parent/Child with Same Percentage Ancestry	144
My Sister is African but I'm Not!	148
A Little Neanderthal in Many People's DNA	155
Tracing Your Ancestry Way Back in Time	159
What Do You Mean Not African?	164
Identifying Richard III and Two Romanovs	170
Mitochondrial Eve and Y-Adam	178

TABLE OF CONTENTS (cont.)

	Page Number
Summary	185
Appendix A: Blood Typing	188
Appendix B: Raw DNA Data into Health Data	198
Image Credits and Links	203

Preface

DNA ancestry and relationship testing have taken off in popularity in the 2010's. Literally millions of people have jumped right into this form of recreational genetic testing to dig into their family's history and find long lost relatives.

And yet, many of these people do not have a firm grasp on what DNA is and how it works. This lack of deeper understanding means they don't get as much as they can out of these tests and that some of the results might be confusing when they don't have to be.

That is where this book will come in handy. Here I explain DNA and how these genetic tests work in easy to understand language so the results make sense.

DNA: Not So Mysterious

When asked about DNA, most people know that with the exception of identical twins, everyone's DNA is unique. And many people know that we share DNA with our relatives.

What most people don't know, though, is why we have a personal set of DNA or how that DNA is passed down. And many biology classes don't help.

Teachers and professors quickly get too complicated when it comes to DNA when there is no need to. The basics are easier to understand than many of these educators make them out to be.

Once mastered, these basics can help you better figure out what that paternity test actually means, how reliable that direct to consumer genetic test might be, why you can't tell from DNA if that person is a half-sibling, nephew or even first cousin and so on.

I have split this book up into three sections. In the first, I go over what DNA is and how it is passed on. In the second part I go over what this means for relationship testing and in the third I go over what it all means for ancestry testing.

There are also plenty of real stories from people about their actual test results.

Let's get started.

Part 1. DNA Basics

You might be tempted to skip over this part and head right for Parts 2 or 3. I urge you not to do this.

This section will use every day examples and analogies to explain DNA to you in a way that is easy to understand. With this better understanding, you will be ready to tackle what DNA tests can and can't tell you and so get the most out of these tests.

This is not like that biology class in high school or college. It is much simpler and, hopefully, more fun.

A Bit About DNA

I won't be talking much in this book about what DNA actually does (maybe I will do that in a future book). Here I want to focus on how DNA is passed down from parent to child and what can happen in the process. This is key to understanding ancestry and relationship DNA test results.

Most everyone has heard that you get half your DNA from your mom and half from your dad. And lots of people know that siblings share half their DNA too.

I have found that while the parent part makes intrinsic sense to people, the siblings seem to make less sense. Many people are confused as to why siblings like brothers and sisters share half their DNA and half-siblings share one quarter of their DNA.

Let's use a deck of cards to explain why siblings share, on average, half their DNA with each other.

Imagine that everyone's DNA is a standard deck of 52 cards. For this example, I'll have the deck of cards be the shared parent's DNA.

So I don't have to keep saying shared parent, I'll say it is mom's DNA. Keep in mind, though, that this works for dad too.

You can do this experiment with a real deck of cards or with a random card generator (I used the one at https://www.random.org/playing-cards.)

First shuffle mom's DNA (the whole deck of 52 cards) and deal out 26 cards. These 26 cards represent the DNA she passed down to her child. It is half of her DNA.

Here is my first set of cards from the random card generator:

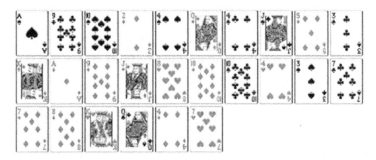

Next you add the cards back into the deck and "reshuffle" to get the second child's DNA, the sibling of the first child. Here is what I got on my second deal:

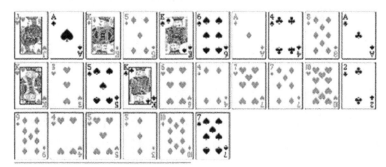

The first thing to notice is that the cards are not all the same. This is the case even though these two siblings each got 26 out of mom's 52 cards.

They didn't get the same 26. They each got half the deck of cards but not the exact same half.

Hopefully this helps explain why two siblings don't share the same 50% of mom's DNA.

Now let's see how many cards these two share in common in this case:

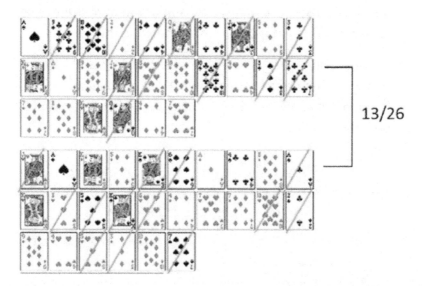

What I've done is put little lines through the cards that don't match. As you can see, these two have 13 of these cards in common.

So in this case, these two siblings share 13 out of 52 of mom's cards. In other words, they share 25% of their mom's DNA. (Note that both are still 50% related to their mom—they each share 26 out of mom's 52 cards.)

If we are dealing with a half-sibling, this is all the two would be related. They'd have different dads and this DNA would be different from the DNA they got from mom. In our analogy, dad might have "DNA" from an UNO deck and so be unrelated to mom.

So these two would share 13 out of 26 from mom and 0 out of 26 from dad. As half-siblings, they would share 13/52 cards or 25%.

In some ways I got very lucky with that first couple of deals. Because the cards are chosen at random, there is no reason I had to get a 13 out of 26 match in this case.

Theoretically there is even a chance that they could all match or that none of them might! Let's see what happens when we are dealt a third hand of 26 (this would be a third sibling):

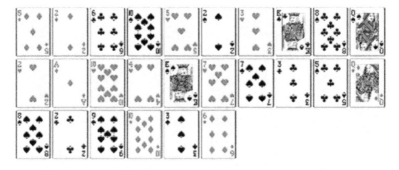

Again this sibling is 50% related to the shared mom...these 26 cards came from her deck of 52. But let's line up all three and see how these three siblings matchup:

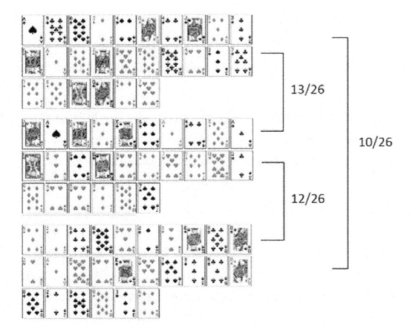

This time I didn't put little lines through the cards that don't line up so you'll have to take my word for the numbers. (Or check to see if I'm right...I may have missed one!)

The first thing to notice is that we didn't get a 13 out of 26 match each time. This is because the 26 that we get is random--as I said, it is theoretically possible to have none or all of them match.

Here we got a 13/26, a 12/26 and even a 10/26. If this were exactly how DNA worked, in the last case the half-siblings would only share about 19% of their DNA. Which, as you'll see a bit later in the book, can and does happen in real life, though very rarely.

This is a very simplified example of how this all works. For example, mom's DNA is made up of way more than 52 "cards." The end result of a larger deck of cards is that we tend to get much closer to the expected percent relationship than we do when we use a standard deck of cards.

So the bottom line is that everyone shares the 50% of their DNA with their mom and 50% with their dad. In other words, sons and daughters are pretty much always 50% related to mom and 50% to dad. But this is where the exact percentages end.

Other relationships just have an average amount that could be shared – it changes case to case. So siblings share around 50% of their DNA, half-siblings around 25% and so on. And once you get out to third cousins, you can have levels that vary a lot from the predicted 0.78% because the little differences can build up with each generation.

I have found that this cards example is a very powerful analogy that helps people understand DNA and how it is passed on. But of course no analogy is perfect. To really understand all of this we need to dive in and take a look at real DNA.

DNA is Packaged in Chromosomes

Most every one of the *trillions* of cells in your body has around 6 feet of DNA. That DNA is split up into 23 pairs called chromosomes. One from each pair of 23 chromosomes comes from mom and one from dad for a total of 46.

Below is what a typical set of chromosomes looks like for a male. I have labeled each chromosome in a pair with either an "M" for mom or a "D" for dad.

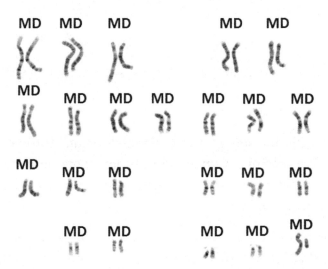

So for each pair, the one on the left came from mom and the one on the right from dad. (In reality, with an image like this we don't know which one came from which parent [except for the last pair]).

This is why we share 50% of our DNA with mom and 50% with dad – we get one from each parent.

You can tell this is a male's DNA because of that last pair that is circled below:

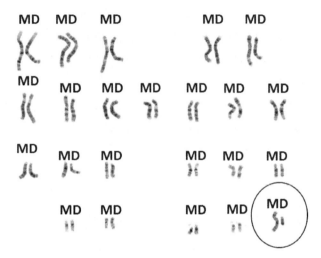

Image: Wikimedia Commons

The big chromosome is the X and the little one the Y. If this were the DNA of a woman, both chromosomes in that last pair would be the same size because she'd most likely have two X's instead.

When we have kids, we pass one from each pair down to each child. This is again why we share half our DNA with our moms and half with our dads.

(This is also why dad determines whether his child will be a boy or a girl. If dad passes a Y, the child will be biologically male and if he passes the X, biologically female.)

Once you get past mom and dad, though, things get a little less precise. This has to do with something that I didn't mention yet that happens before we pass our DNA down to our kids—recombination.

Recombination is when DNA is swapped between two chromosomes in a pair. So for example, before mom passes down a chromosome 1 to her child, her two 1's swap DNA between each other. The child then gets the new mash-up of chromosome 1.

The end result of this is that mom doesn't give you one chromosome or the other from each pair. She gives you a mix of the two. Same thing with dad.

It is this mixing that can cause the wide range of shared DNA between similar relatives.

Let's do a quick example following a single pair of chromosomes. Remember, what I am describing happens for all 23 pairs. (Well actually only 22 pairs in men. As you'll see later, the X and Y do not recombine.)

Let's say these are a pair of mom and dad's chromosomes:

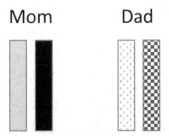

I have drawn mom's in gray and black and dad's with two different patterns so we can follow them more easily. Let's imagine they have two kids:

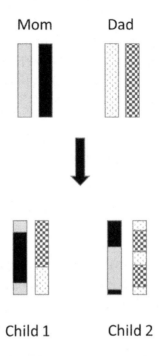

As you can see, neither mom nor dad passed on one of their chromosomes down completely—neither child, for example, got an all gray or all black chromosome from mom. Instead mom passed on a mix of gray and black and dad passed on a mix of his two patterns.

The other thing you probably noticed is how different the DNA is between child 1 and child 2. They each came from the same parents but have different mixes of each parent's DNA. This is a big reason why you can be so different from your brother and sister!

Let's look at this image a little more closely:

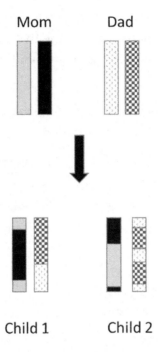

Notice that the first chromosome of Child 1 is a mix of mom's gray and black chromosomes. The top and bottom are from the gray one and the middle is from the black one.

For child 2 the chromosome from mom is a different mix. It is black at the top with most of the rest being gray. There is a sliver of black at the bottom.

This is a consequence of DNA swapping being pretty random where it happens along the chromosome. So for child 1's chromosome from dad there was just one spot where the DNA swapped meaning the top half came from the one chromosome and the bottom from the other one.

Child 2's chromosome from dad came from two separate pieces being swapped between the two patterns. That is why there are two checkered bands and three speckled ones.

OK so that is why siblings can end up with different DNA combinations compared to their parents.

Now let's dig a little deeper and see why this recombination can lead to different amounts of shared DNA between these siblings. In other words, I'll try to explain why siblings usually share a bit more or a bit less than half their DNA.

Around 50%

So here again are the two siblings we are talking about, Child 1 and Child 2:

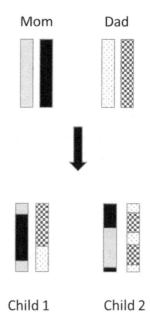

To make it easier to see, I am going to line up the chromosome Child 1 got from mom with the one Child 2 got from mom. And do the same with the chromosomes they got from dad:

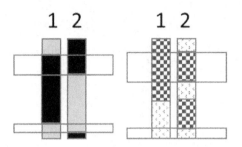

So the chromosomes from mom, the ones with black and gray, are on the left and the ones from dad are on the right. The first of each pair came from Child 1 and the second from Child 2.

Those open rectangles I added are the parts of the DNA that the two share. For the chromosomes from mom, Child 1 and Child 2 share some black up near the top and a bit of gray near the bottom and for the ones from dad they share a big checkered block near the top and a speckled bit near the bottom.

In this case it looks like these siblings share a bit less than 50% of their DNA but remember, there are 22 other pairs all doing the same thing. The amount of shared DNA will even out to around 50% but it almost certainly will not be an exact number. Different siblings will share a bit more or a bit less of their DNA.

But you can probably see why it isn't an exact thing. They might by chance have ended up sharing more or less DNA.

Or even all or none at all. (The chances for this last possibility are so remote that it would essentially never happen in real life.)

Now let's have these two siblings have kids:

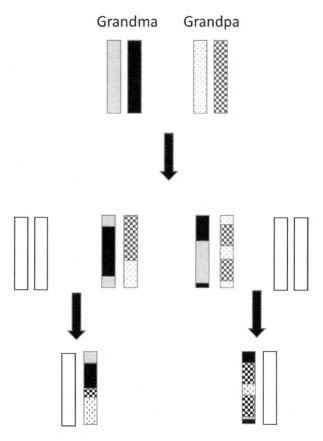

I am starting out with the original parents whom I have renamed grandma and grandpa and the original children are in the second level. I have used empty rectangles for their spouse's DNA because it doesn't matter for our discussion here.

Unless they happen to be related, the spouses won't share any DNA from recent relatives. They won't have any gray, black, speckled, or checkered DNA.

The third level shows the resulting first cousins. Take a close look and you can see that by chance these first cousins share no DNA on this particular chromosome.

Of course they will probably share more on their other 22 pairs but the total may not be the 12.5% number usually quoted for how related first cousins are.

There are cases where first cousins share only 2% of their DNA instead of the theoretical 12.5%. And cases of 9% or 15% are quite common.

There is an online site, ISOGG Wiki (http://isogg.org/wiki/Autosomal_DNA_statistics), that has looked at the actual DNA results of people of known relationships to see how much DNA various relatives actually share. The ranges are surprisingly large.

For example, when they looked at 662 first cousins, instead of always finding the predicted 12.5% shared DNA, they found a range 1.2-22.9% of shared DNA.

A picture is more powerful so here is what those numbers look like in a pie chart:

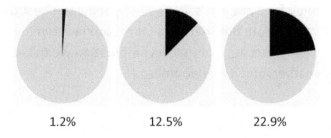

1.2%　　　　　　12.5%　　　　　　22.9%

The black slice is the amount of shared DNA. As you can see, it is pretty amazing the range that can be shared between first cousins!

The site looked at lots of other relatives too. For example, they found a range of shared DNA of 12.9-33.8% for 195 pairs of grandparents and grandchildren. Here is what that looks like in a pie chart:

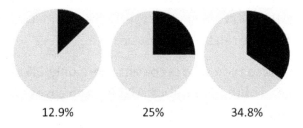

12.9% 25% 34.8%

Since the ranges of shared DNA between first cousins and grandparent/grandchildren overlap, it is possible for a grandchild to look more like a first cousin in terms of shared DNA. Which means a test could conceivably miscall a grandchild as a first cousin.

And of course a test can miscall a half sibling as a grandparent/grandchild as they are both predicted to share 25% of their DNA. Embarrassingly enough, this is exactly what happened with my half-sister and me!

This is why the relationship a company assigns you should be taken with a grain of salt. They are doing the best they can with the data but there is a lot of slop in assigning relationships because of that recombination we talked about earlier, and its randomness.

And why it gets harder and harder to assign a relationship the less related two people are. The more distant the relationship, the wider the possible ranges.

Exceptions: Mitochondrial DNA, X, and Y

So that is how most of our DNA is passed on. The DNA between chromosomes in a pair swap DNA and then the resulting mixed chromosomes are put into the sperm or egg.

There are two bits of DNA that (mostly) don't recombine. The first is mitochondrial DNA (mtDNA) and the second is the X and the Y chromosomes in men.

Mitochondrial DNA (mtDNA). The mtDNA is a cool relic from our evolutionary past. Instead of just two copies per cell like our chromosomes, our cells have many more copies of mtDNA.

The other big difference is where this DNA is stored in a cell. Most of our DNA is stored in a part of the cell called the nucleus, the blob in the image below.

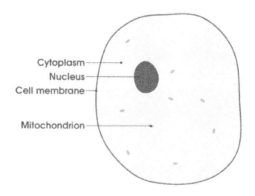

Image: Wikimedia Commons

But as you might guess from its name, mtDNA is stored in the mitochondria (the specks above). It is a separate bit of DNA.

It is also different in that it passes from mom to kids; in fact, dad's is destroyed right after an egg is fertilized! This means that your mtDNA only came from your mom.

It is pretty weird that mitochondria have their own DNA. After all, a mitochondrion's job is to make energy for the cell. Why would the cell's "powerhouse" have its own DNA?

Because it used to be a creature living on its own. Or at least that is what scientists believe.

Scientists think that a long time ago, our ancestors absorbed the mitochondrion's ancestor. Working together, this duo took over the world.

But in the process, the mtDNA took a beating. Most of this DNA is now found in the nucleus of the host cell (although different species have mitochondria with differing amounts of DNA). The mitochondria can no longer live on its own -- it is a hollow shell of what it once was.

And this hollow shell passes only from mother to child. Of course dad has mitochondria too, but for a couple of reasons kids almost never get them from him.

The first reason is that the egg is way bigger than the sperm meaning that it has a whole lot more mitochondria. This means that when the egg and sperm combine, the egg's mitochondria swamp out the sperm's.

But this isn't good enough for the egg. The egg actually marks sperm mitochondria for destruction and gets rid of them. The end result of all of this is that we get our mtDNA exclusively from our mothers.

The fact that you get only mtDNA from your mother will become very important for the ancestry section of the book.

X and the Y from Dad. The other exception to how DNA is passed down is the X and the Y chromosomes in men.

Because the X and the Y are so different, they don't swap much DNA. In other words, they don't recombine.

What this means is that daughters get dad's X and sons get dad's Y. This is different than the X both get from mom.

Like the rest of our chromosomes, mom's two X's swap DNA before she passes an X to her child. Both sons and daughters end up with a mix of both of mom's X's.

This is why you will sometimes see stories about grandmothers and how much DNA they share with their grandkids.

Shared DNA and Grandmas

Grandmas are predicted to share a bit more or less DNA with their grandkids depending on how she is related to them. For example, a grandmother is more likely to share a little bit more than 25% of her DNA with the daughter of her son. Instead, she shares on average about 26% of her DNA.*

The reason has to do with how dads pass on their X and Y chromosomes to their kids. Instead of passing down a mix of these two chromosomes, dad passes an unchanged X to his daughters and an unchanged Y to his sons.

As you'll see below, this means a paternal granddaughter ends up getting an X chromosome that is made up almost exclusively from grandmother's two X's. This lets this granddaughter share a little extra DNA with that grandmother.

To see how, we are going to look at a pair of chromosomes being passed down again. Keep in mind that there are 22 other pairs of chromosomes that we aren't showing!

As you might recall, before a chromosome gets passed down to a child, the two chromosomes in each pair swap DNA. As you can see in the image on the next page, the result is two, brand new, unique chromosomes. Again, this is called recombination.

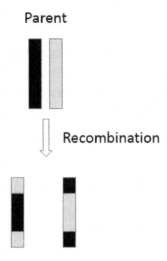

One of these mixed up chromosomes is then randomly passed down to a child. Below, see what happens when recombination happens in a dad and he has kids.

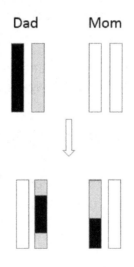

Both children have 50% Mom DNA and 50% Dad DNA

(I'm not showing DNA swapping in the mom even though it happens. Also note that this is just one way the chromosomes might have swapped their DNA. There are nearly an infinite number of possible combinations!)

Since Dad passes down only one chromosome to his child, he shares 50% of his DNA with each of his children. The same goes for mom. And a child shares 50% of his or her DNA with each of his or her parents.

For simplicity, let's name the dad John. Where did John get each of his chromosomes? He got one from his mom, and one from his dad, as you can see below.

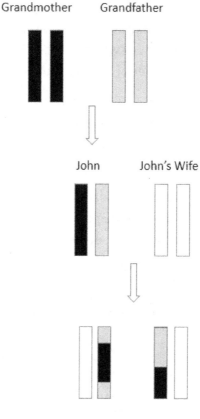

Each child has 25% of their
DNA from the grandmother

(You will see that I didn't show the chromosomes of the grandparents, or of John's wife, swapping DNA. They actually do — but that would make a more complicated picture. So just keep in mind that they do swap DNA in reality.)

Back to the picture. The grandmother shares half of her DNA with John. And John's children each get only half of their DNA from him. So this means that the grandmother shares a half of a half of her DNA with each of her grandchildren – or 25%. Just those blocks of black in the picture.

This is why grandparents share on average 25% of their DNA with their grandchildren. The exact amount depends on how much DNA is swapped. But when you look at all 22 pairs of chromosomes (not just the one pair above), it usually averages out to around 25%.

If this is all there was to it, grandmothers would always share about 25% of their DNA with their grandkids. But we know that some granddaughters are a bit more related to their dad's mom.

The difference comes from the pair of chromosomes that don't swap much DNA. These are the X and the Y chromosomes, or the sex chromosomes, in men.

If you have two X chromosomes, you are almost always biologically a female. If you have one X and one Y chromosome, you are almost always a male biologically. (Males get a Y from dad and an X from mom.)

DNA can only be swapped between matching chromosomes. Two chromosome 1's can swap DNA, and so can two chromosome 2's, 3's, etc.

A woman's two X chromosomes also swap DNA. But a chromosome 1 won't usually swap with a chromosome 3. And because they are also different, a man's X and Y don't either.

Take a look at what happens now when we focus on John's X and Y chromosomes:

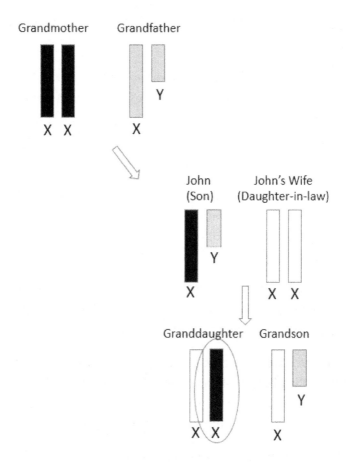

(To keep it simple, we're still not showing the DNA swapping that happens in John's wife and in the grandparents.)

John's daughter actually gets John's X chromosome completely unchanged. And it is the same X that John got from the grandmother. This is what makes the granddaughter and the grandmother more likely to share a bit more DNA!

DNA sharing for all the other 22 chromosome pairs is around 25%, as we showed before. But for the sex chromosomes shown above, there's a little bit more sharing.

John's daughter has a bit more of the grandmother's DNA and a little bit less of the grandfather's! And as you can see, the grandmother shares a little bit less DNA with John's son.

To make it even clearer, we can compare this to what happens if a grandmother and grandfather had a daughter (with two X chromosomes) instead. As you can see below, DNA can swap between the daughter's X chromosomes.

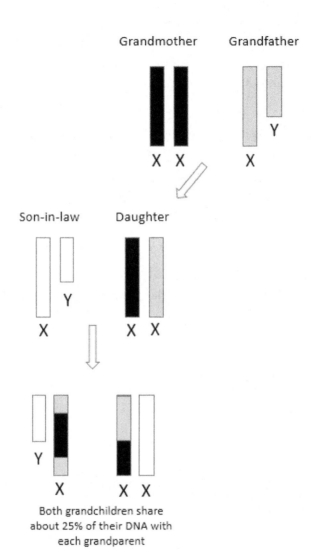

Both grandchildren share about 25% of their DNA with each grandparent

The children of the daughter share the normal 25% of DNA on the sex chromosomes with their grandparents. This happens for their other 22 pairs of chromosomes too. So a maternal granddaughter and grandson are equally related to grandma.

Let's put the whole family together now. We'll show recombination happening in the grandparents and daughter-in-law, too.

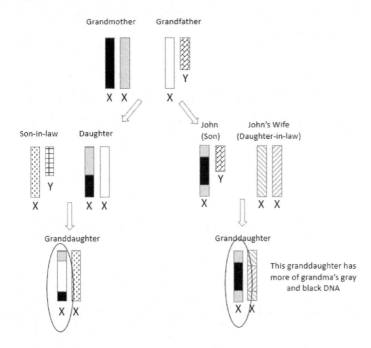

You can see that John's daughter has a little bit more of the grandmother's DNA on the X chromosome than does the maternal granddaughter. (Just a bit more gray and black.) So John's daughter shares 26% of the grandmother's DNA — but the other granddaughter shares the normal 25%.

*If you're interested, here's the rough calculation for how it is that 26% of DNA is shared between the grandmother and granddaughter. The grandmother and granddaughter share 25% of the non-sex chromosomes, and there are 44 of them (22 chromosome pairs). This adds up to 11 *chromosomes shared. But, they also share their X chromosome. This is a total of 12 chromosomes. Out of 46 chromosomes (23 chromosome pairs), this comes out to 26%. (The sizes of the chromosomes were not included in the calculation.)*

Now you have enough information to start digging into genetic tests and what they can and can't tell you. And why.

About Genetic Tests

Genetic tests are becoming both more popular and more commonplace. Unfortunately many if not most people don't have the background needed to choose the right test and to understand and interpret the results.

Hopefully the previous section gave you some of this needed background so you can apply it to whatever DNA test result you get.

For example, you'll understand why the test you get to see if your dad is really your dad is not very useful for telling if your uncle is really your uncle even though many companies sell the same or a similar test for both. And why an ancestry test that looks at which African tribe your ancestors might have come from can turn out to not be useful. For example, one kind of this test might show former US President Obama as non-African!

These are just two of many possible problems. This book should help you find the best test available for your particular problem and help you sort out which results matter and which ones you can ignore. In addition, you'll get a feel for the limits of these tests and begin to understand how the results of the test could be wrong.

Part 2: Relationship Testing

As the paternity episodes on the TV show *Jerry Springer* highlight, a common reason people turn to DNA testing is to figure out who is the father of a child.

Nowadays people usually use a DNA test but the first clues that a man may not be the dad can come from blood typing. For example, if a potential dad is AB and the child is O, odds are that this man is not the father (see Appendix A to see why this is and for some exceptions to this rule.)

In fact, before DNA testing blood typing was a common way to rule out that a man was the dad.

While pretty good at showing that a man is probably NOT the dad, it isn't very good at showing that a man IS the dad. A match means he is one of many men who could be the father.

And sometimes a blood test result that usually means a man couldn't be the father is wrong. On rare occasions, a seemingly impossible blood type result is actually possible.

For those interested readers, I have included a whole section, Appendix A, that deals with the genetics of blood type and how even these results can be wrong in a paternity test.

Now onto the DNA tests!

DNA Relationship Test Advice

Some things to keep in mind when thinking about which test to get and how to interpret the results:

Paternity Tests

1) If you get a paternity test, don't just focus on the testing company's conclusions. Take a hard look at the results and see how many places in the DNA that you and your potential child do not match. If there are two or three of these "mismatches" and the testing company concludes that you are not the father, you might want to consider additional testing.

2) Remember to let the company know if potential dads are related. If they are, try to have both men tested but if that isn't possible, make sure to do additional testing. You may even want to consider using one of the more powerful relationship tests offered by companies like 23andMe, Ancestry.com or Family Tree DNA.

3) Paternity tests are not foolproof. There are bad companies that do sloppy work and there can be human error even with the best companies. If the results don't make sense, think about getting a second test from a different company. At a minimum look for a company that is AABB accredited.

This link lists AABB accredited companies: http://www.aabb.org/sa/facilities/Pages/RTestAccrFac.aspx

4) If you have had a bone marrow transplant, let the testing company know. The donor's DNA can sometimes interfere with the test results.

5) There are rare cases where every DNA test comes up with a negative result and the man is still the father. We give a couple of these examples.

Relationship Tests

1) If you are interested in using DNA to figure out a relationship more distant that parent/child, you should go with the more powerful relationship tests offered by companies like 23andMe, Ancestry.com and Family Tree DNA. They will conclusively show two people are related up to second cousins and have a pretty good shot at seeing even more distant relationships.

2) While these tests are great at showing people are related, they are not always accurate in assigning the relationship. This is not the test's fault but instead has to do with how our biology works (see the first section of the book for more on this).

3) Keep in mind that as you figure out your family tree, there may be a few unwanted surprises in there. Many a family has realized that a sibling like a brother or sister is a half instead of a full sibling or that he or she was adopted.

4) There are good online tools like GEDmatch that can help you get more out of your data.

5) Having your mitochondrial and Y DNA tested can be useful for getting the most out of your relationship data.

6) If you have had a bone marrow transplant, let the testing company know. The donor's DNA can sometimes interfere with the test results.

7) There are rare cases where every DNA test comes up with a negative result and the man is still the father. We give a couple of these examples.

8) Be careful about using online sites like Promethease to pull health data out of your raw data. Because we do not yet have a good grasp on how our DNA impacts our health (a future book perhaps), it is hard to assess how important a lot of the data is. Also keep in mind that you may end up getting different health data using different company's test results because they do not all look at the same DNA. More about this in Appendix B.

"Simple" Paternity Tests

Advice: Do not just rely on the testing company's conclusions. Really look over your results and see if they came to their conclusion because you and the potential child did not share DNA at two or three places (a quick and easy way to do this is by looking for how many markers have a paternity index of 0.) If there are two (or maybe even three) mismatches and the test claims he is not the dad, think about getting additional testing.

DNA paternity tests rely on the fact that we get half our DNA from our mom and half from our dad. They also focus on parts of our DNA that are less likely to be shared by any two people. This last part is harder than you might think since any two people share around 99.5% of the same DNA!

Let's see how this all works in practice.

A paternity test compares a small part of a child's DNA to a small part of a potential dad's DNA. In addition, the test is made much stronger by looking at these spots on mom's DNA too.

A common test looks at 15 or so of these spots that go by lots of different names. They're called microsatellites, simple sequence repeats (SSRs), short tandem repeats (STRs), or variable number tandem repeats (VNTRs).

These are places in the DNA where a certain bit of DNA is repeated. For example, some people may have 10 repeats of a small bit of DNA at a certain site while others might have 9 or 11 or whatever.

These results are reported back to you for each spot they look at. It might look something like this:

D3S1358, 17/18.

So this means that at spot D3S1358, you got 17 repeats from one parent and 18 from the other.

(As an aside, many people are confused by the fact that these tests will sometimes report just one number. This does not mean the tested person got DNA from just one parent. Instead, it means that they got the same bit of DNA from both parents. For technical reasons they report a 17/17 result as 17.)

Let's look at a single marker to see how this kind of test works. Imagine that mother, son, and potential dad have the following markers at D3S1358:

Mother: 13, 14

Son: 13, 18

Potential Dad: 17, 18

So mom has a 13 and a 14, the potential dad a 17 and an 18, and the son a 13 and an 18.

Here is another way to show this:

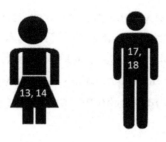

We know that mom is a parent so first we look at which STR she might have passed to her son. The only possible one is the 13 I have highlighted below (mom does not have an 18).

Mother: 13, 14

Son: 13, 18

Potential Dad: 17, 18

This means that his father had to contribute an 18. And yes, this man, the potential dad, has an 18 to give. So he could be the dad!

Now lots of men have at least one 18 at D3S1358 which is why you need to look at more markers. Statistically if dad matches on 15 out of 15 of these markers, odds are he is the dad. As you'll see below, not always. But usually.

And you might think the converse is true. If he has mismatches he is not the dad. This is not always true.

It turns out that many dads don't match at every tested spot even though they are the dad. This is so common that the chances of a mismatch are baked into the test results. A mismatch at one or two markers is usually not enough to exclude a man as a father. Some companies even allow three mismatches!

This is confusing to some people because if the child got the DNA from dad, why doesn't it match? Turns out it is because the DNA changed on the way from father to son. The DNA got a mutation.

Mutations Are More Common Than You Think

Imagine we got this result for our paternity test at the D3S1358 marker:

Mother: 13, 14

Son: 13, 18

Potential Dad: 17

From this result we can see that the son got a 13 from mom which means he needs to get an 18 from dad. But dad has only 17's to offer! (Remember, these companies present the results as a single number if they are the same and so 17 really means 17, 17.)

This would seem to rule out our potential dad as the father of this boy. And it does unless his 17 changed to an 18 on the way to being passed to the child. Which happens more often than you might think.

Because testing companies don't look at a lot of DNA, they need to pick DNA that tends to be different between people. These STRs fit the bill perfectly. The odds are strongly against any two random people matching up at all 15 of these spots.

Unfortunately what makes them so useful can also cause problems with paternity test results. This is because these particular spots on our DNA have a higher chance of changing between generations than many other spots in our DNA.

The reason these DNAs tend to change or mutate more often than other DNAs is because they are DNA repeats.

They are repeats of the same bit of DNA over and over again.

It turns out that the cellular machinery that copies DNA can't always copy this repeated DNA exactly right.

As it is copying a stretch with repeats, this machinery (called a DNA polymerase) will sometimes add a repeat or two or take one or two away. From a health perspective this isn't a big deal as these bits of DNA don't seem to be doing a lot in the cell. But it can wreak havoc with test results.

In fact as I said it is common enough for there to be a single difference between father and son that this possibility is baked into the test results. In other words, with most companies if the results are different at one marker, the company will call the man the father.

But there is less consistency between companies when the number of mismatches go beyond one. Some will tolerate two or even three.

What this all means is you want to look at your test results very closely. If a testing company concludes that a man is not the father based on two mismatches, keep in mind that he still could very well be the dad. Two mismatches are much more common than you might think.

This is why you should always look at more than the conclusion the company gives you. And why a deeper understanding of our DNA and how these tests work can be so important.

False Positives: When You Look Like the Dad But Aren't

Advice: Make sure your testing company is AABB accredited.

Many people place a lot of trust in DNA paternity tests. Perhaps too much.

It is always important to keep in mind that like anything else, these tests can be wrong. For example, there is always the possibility of human error—the sample could have been mixed up at the lab or the lab could have done a sloppy job of running the test.

This is why it is critical that you get your test from a reputable company. Unfortunately it is often difficult to tell the reliable companies from the shady ones.

Your best bet is to make sure that the company has the right accreditation. At a minimum look for AABB accreditation. Maybe something like this:

If a company is not accredited by AABB, do not use them.

But testing error isn't the only potential problem. Sometimes the test can be run properly and still be interpreted incorrectly and so give the wrong result. We'll go over a few of the ways this might happen.

This link lists AABB accredited companies:
http://www.aabb.org/sa/facilities/Pages/RTestAccrFac.aspx

My Brother is the Dad

Advice: If you want a good result, be up front with the testing company if the two men who might be the fathers are related. And if at all possible, have both tested.

There are some situations where a man looks like he is the dad even when he isn't. A common way for these sorts of false positive results to happen is when the real dad is closely related to the man who is tested.

Imagine two brothers who might be the father of a child. For whatever reason, no one tells the paternity testing company that this is the case and so only one brother ends up getting tested.

The results show that the tested brother's DNA matches the child's and so he is identified as the dad. Except that he isn't. His brother also happens to match the child in this DNA test.

What needed to happen here was for someone to tell the testing company upfront that either brother might be the dad. Then the company could test both of them and do additional tests until one or the other brother was ruled out.

And of course for this to happen they would need to have picked a good company that is willing to take the time to give this level of analysis. Not every company will.

At first this might seem weird that a father's brother could be mistaken for the father since, unless the brothers are identical twins, they don't have the exact same DNA. You'd think this would mean that a DNA test would never confuse two people.

And you'd be right if DNA paternity tests looked at all of someone's DNA. But they don't.

As I said before, current DNA paternity tests only look at a small fraction of anyone's DNA. This is usually enough to tell which of two *unrelated* men is the father but it doesn't always work for related men.

Related people share a lot of the same DNA (this is why paternity tests work in the first place!). This makes them more likely to share the bits of DNA that get tested too. And this makes it more likely that they will be hard to tell apart on some paternity tests.

This isn't just a theoretical problem, either. A friend of mine at a paternity testing company tells me that these issues really do come up. And the only way around it is to look at more spots on the DNA.

The number of markers that testing companies look at works great for unrelated men. It is very unlikely that two random men will match up at all 12, 15, or 20 markers.

But brothers aren't two random men. They have the same mom and dad and so share around 50% of their DNA.

What this means is that they are more likely to share at least one marker at many of the tested sites. And therefore they are more likely to each pass the same DNA markers to their kids.

This is where the trouble can start. If two brothers happen to share markers at all of the sites tested, then the wrong man can be identified as the dad.

But again, unless they are identical twins, the brothers will not match up at every DNA position. If a company looks at enough DNA, they will be able to tell who the dad is. And a company will do this if you tell them that either brother might be the dad.

Let's use an example to show how two brothers might be indistinguishable as the dad and how testing more DNA might resolve the problem. To simplify things, I'll start off with just 10 markers. Here is a test result for mom, son, brother 1 and brother 2:

Marker	Mom	Son	Brother 1	Brother 2
A	**13,15**	**13**, *13*	*13, 13*	*13, 13*
B	**9, 12**	**9, 12**	8, 12	12, 17
C	**11, 13**	**11**, *14*	*14, 14*	*14, 14*
D	**29, 31**	27, **31**	25, 27	27, 27
E	**16, 21**	**16**, *22*	21, *22*	*22*, 23
F	**14, 17**	**14**, *19*	16, *19*	*19, 19*
G	**16, 28**	**16**, *19*	*19*, 28	*19*, 25
H	**12, 12**	**12**, *15*	12, *15*	*15, 15*
I	**7, 9**	**7, 9**	*6, 7*	*6, 7*
J	**8, 8**	**8, 8**	*8, 8*	*8, 8*

In this example, the shared markers from mom are **bolded** and the shared markers from the potential dads are *italicized*. (Other markers are left in normal text.)

Notice that once we figure out which markers came from mom (**bold**), either brother could have given the child the second marker (*italicized*).

As an example, here is marker A:

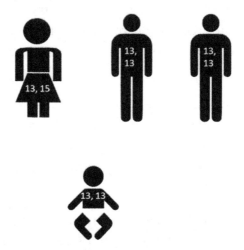

The baby got one of his 13's from mom and so needs to have gotten his second 13 from dad. Both dads have a 13 to give so either one could be the dad. Look over the table on the previous page and you'll see this is true of all 10 markers.

If just brother 1 were tested, then he would be identified as the dad even though he might not be. (The same is true for brother 2.)

Luckily for everyone involved, both brothers were tested. Since the first batch of results was inconclusive, more extensive testing is done. The results are on the next page:

Marker	Mom	Son	Brother 1	Brother 2
A	**13,**15	**13**, *13*	*13, 13*	*13, 13*
B	**9,** 12	**9,** *12*	*8, 12*	*12, 17*
C	**11,** 13	**11,** *14*	*14, 14*	*14, 14*
D	29, **31**	*27*, **31**	*25, 27*	*27, 27*
E	**16,** 21	**16,** *22*	*21, 22*	*22, 23*
F	**14,** 17	**14,** *19*	*16, 19*	*19, 19*
G	**16,** 28	**16,** *19*	*19, 28*	*19, 25*
H	**12, 12**	**12,** *15*	*12, 15*	*15, 15*
I	7, **9**	7, **9**	*6, 7*	*6, 7*
J	**8, 8**	**8,** *8*	*8, 8*	*8, 8*
K	**13,** 17	**13,** *19*	*15, 19*	*19, 19*
L	6, **8**	*3*, **8**	*3, 7*	*3, 3*
M	**15,** 18	*2*, **15**	*3, 3*	*2, 3*
N	**17,** 19	**17,** *18*	*9, 18*	*7, 18*
O	18, **22**	*18*, **22**	*18, 18*	*18, 18*

When we add five more markers, brother 1 now has a mismatch (the gray box). He has two 3's at this marker while his brother has a 2 and a 3. The child got a 15 from mom which means the child had to get a 2 from dad. Only brother 2 has a 2 to give

Now if we don't test brother 2 as well, this still isn't enough to exclude brother 1. Remember, one mismatch is OK because of the mutating DNA we talked about earlier.

This means we still need some additional testing to find the real dad:

Marker	Mom	Son	Brother 1	Brother 2
A	**13,**15	**13**, *13*	*13, 13*	*13, 13*
B	**9,** 12	**9**, *12*	*8, 12*	*12, 17*
C	**11,** 13	**11**, *14*	*14, 14*	*14, 14*
D	29, **31**	*27,* **31**	*25, 27*	*27, 27*
E	**16,** 21	**16**, *22*	*21, 22*	*22, 23*
F	**14,** 17	**14**, *19*	*16, 19*	*19, 19*
G	**16,** 28	**16**, *19*	*19, 28*	*19, 25*
H	**12, 12**	**12**, *15*	*12, 15*	*15, 15*
I	7, **9**	*7,* **9**	*6, 7*	*6, 7*
J	**8, 8**	**8**, *8*	*8, 8*	*8, 8*
K	**13,** 17	**13**, *19*	*15, 19*	*19, 19*
L	6, **8**	*3,* **8**	*3, 7*	*3, 3*
M	15, **18**	*2,* **15**	*3, 3*	*2, 3*
N	**17,** 19	**17**, *18*	*9, 18*	*7, 18*
O	18, **22**	*18,* **22**	*18, 18*	*18, 18*
P	**3,** 7	**3**, *12*	*5, 8*	*8, 12*
Q	26, **28**	*9,* **28**	*9, 9*	*9, 9*
R	**5,** 6	*4,* **5**	*6, 12*	*4, 12*
S	**9,** 10	**9**, *12*	*12, 18*	*12, 17*
T	**13,** 22	**13**, *19*	*17, 23*	*17, 19*

Now we're getting some results that can tell these two men apart.

Brother 1 has 4 mismatches at markers M, P, R, and T while brother 2 has none. Almost certainly brother 2 is the dad but they may want to do some additional testing to really nail it down.

If the first test on markers A-J had been done on just brother 1, he would be identified as the dad. And might be raising his nephew as his son!

Luckily everyone was honest with a very good testing company. Together they were able to figure everything out.

As a final note, 20 markers may not even be enough. I know of a case where two brothers were identical at 21 of 23 markers and who shared a marker at the other two.

In this case, if the testing company looked at the identical 21, they wouldn't be able to distinguish between these two men as fathers. And even if they looked at 23, there would be a 1 in 4 chance that you still couldn't tell who the dad was even with 23 markers!

The bottom line from both mutations and related dads is that the more DNA tested the better. This becomes really clear when companies try to use these tests for relationships other than parent/child.

Relationships other than Parent/Child

Advice: If you want to figure out a relationship other than father/child, turn to a test like the ones offered by 23andMe, Ancestry.com and Family Tree DNA. You are much more likely to get a meaningful result compared to other companies' tests.

In the last section we showed how paternity testing might not be able to tell which brother is the father of a child. These tests simply do not look at enough DNA to get the right answer every time.

This becomes a real problem when people try to use these tests to figure out more distant relationships. And we aren't talking fifth cousins here. Paternity-type tests can have real problems telling a brother and sister are related!

So you definitely want to be careful when picking out a test to try to figure out if you and your potential sibling share one or both parents. For some tests this is an easy thing to see while others can't give an answer for sure. These second tests can only report back the odds that the two of you are related as half or full siblings.

The tests to be careful with are the ones called siblingship, avuncular, grandparentage and so on. For the most part these tests use the same sort of testing found in paternity tests and they often don't look at enough DNA to reliably and conclusively see relationships like uncle/nephew, grandparent/grandchild and so on.

For these sorts of more distant relationships, you want to turn to a test that looks at hundreds of thousands of markers instead of just tens of them. A variety of companies, like 23andMe, Ancestry.com, and Family Tree DNA (to name just three), sell such tests.

These more powerful tests are often not that much more money. And they have a much higher chance of giving you a meaningful result. Much, much higher.

In the next couple of sections I'll go over in more detail why the paternity-based tests aren't usually powerful enough to conclusively see these relationships. I'll also show you why the other tests work so much better.

Paternity Based vs. Other Relationship Tests

As I have said, sometimes people want to use a genetic test to tell if they are related as brothers, cousins and so on. Many companies will try to sell you the kind of test we've talked about so far that tries to show relationships other than parent/child. Very rarely will they give a definitive answer.

This has, again, to do again with how genes are passed on.

We do all share 50% of our DNA with our moms and 50% with our dads. And on average we share 50% with our siblings.

And yet, even though both are 50%, a paternity type test is very good at telling if a man is a father of a child but not nearly so good at telling if he is a brother of another man. And even worse at seeing if he is the uncle to a child.

Again we'll use an example to show why this is.

Imagine we have two people, Stan and Pierce, who want to figure out if they are full or half siblings. Their parents are not available for testing. There are only the two of them.

Let's say they get a siblingship DNA test. Here are part of their results:

Marker	Stan	Pierce
A	11, 11	12, 12
B	18, 20	17, 20
C	5, 20	5, 17
D	11, 18	11, 17
E	9, 10	8, 11
F	12, 12	12, 15
G	9, 13	8, 12
H	10, 15	10, 15
I	12, 20	13, 20
J	14, 16	13, 15
K	30, 32	29, 30
L	24, 25	23, 25
M	5, 9.3	6, 7
N	8, 10	8, 10
O	17, 17	17, 19

In the left most column we have the names of the bits of DNA being tested. Then we have Stan and Pierce's results at each position.

So, for example, at the first one, A, Stan got an 11 from one parent and an 11 from the other. (I have written it this way instead of the single 11 you'd actually get.) And Pierce got a 12 from each parent.

Looking at all of these results, it is hard to know either way. Are they full or half-brothers? Are they related at all?

It is very difficult to tell from so little data.

Remember, half siblings share 25% of their DNA and full siblings share 50%. But the test is only looking at the tiniest sampling of each person's DNA.

Such a DNA testing company would do its best and use various statistical methods to try to figure it out. And because it is so tricky, they would only be able to tell them how likely they are to be related as full or half-brothers. Stan and Pierce would probably not get a solid answer.

If we add mom and dad it becomes clear how they are related:

Marker	Stan's dad	Stan's mom	Stan	Pierce
CSF1PO	*11, 12*	**11, 12**	**11,** *11*	*12, 12*
D2S1338	*17, 18*	**15, 20**	*18,* **20**	*17,* **20**
D3S1358	*5, 5*	**17, 20**	*5,* **20**	*5,* **17**
D5S818	*17, 18*	**11, 11**	*18,* **11**	*17,* **11**
D7S820	*9, 11*	**8, 10**	*9,* **10**	*11,* **8**
D8S1179	*12, 15*	**12, 14**	*12, 12*	*12, 15*
D13S317	*8, 9*	**12, 13**	*9,* **13**	*8,* **12**
D16S539	*15, 15*	**10, 20**	*15,* **10**	*15,* **10**
D18S51	*20, 20*	**12, 13**	*20,* **12**	*20,* **13**
D19S433	*13, 16*	**14, 15**	*16,* **14**	*13,* **15**
D21S11	*29, 32*	**30, 31**	*32,* **30**	*29,* **30**
FGA	*23, 24*	**20, 25**	*24,* **25**	*23,* **25**
THO1	*5, 7*	**6, 9.3**	*5,* **9.3**	*6, 7*
TPOX	*10, 10*	**8, 11**	*8, 10*	*8, 10*
vWA	*17, 17*	**17, 19**	*17, 17*	*17,* **19**

Here dad's markers are *italicized* and mom's are **bolded**.

Stan shares markers with both of his parents and Pierce shares markers with both of Stan's parents as well. This strongly suggests they are full brothers but again it is not a for sure thing. While unlikely, it is still possible they are half siblings, especially if Stan's dad and Pierce's real dad were related.

To get a better result Stan and Pierce should use a test like the one offered by 23andMe and other similar companies. With the right test from these companies, it should be obvious whether they are full or half-brothers. Or completely unrelated.

The difference between these tests and the less powerful ones is how much DNA they look at. The first set looks at under 100 different spots on the DNA while the second set looks at hundreds of thousands of them.

By looking at so many different parts of the DNA, the more powerful tests can actually say whether or not two people share enough DNA to be half siblings. The other tests can almost never do this.

The reason for this isn't as obvious as you think. It isn't just that you have more markers which makes the statistics better. Instead, by looking at so much of the DNA, the more powerful tests can actually reconstruct both people's DNA.

They are now essentially comparing all of the DNA of these two people instead of just a few spots. This lets them see how related two people are.

On the next page you can see three examples of the kind of data you can get from 23andMe. Don't worry right now about understanding what they are comparing or what the data means. For now, just look them over and see how they differ.

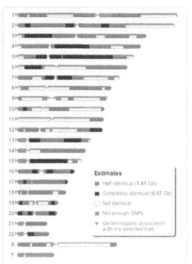

HALF-BROTHER   FULL BROTHER

UNRELATED

(Side note: These are not what the test results currently look like but they are still as easy to distinguish. Also they

are usually dark blue [completely identical], light blue [half-identical], white [not identical] and gray [not enough information]. Here the image is shown in black and white which makes the light blue and the gray look similar.)

A quick look and you can see pretty easily that these three are easy to tell apart. The three patterns are very different.

Half siblings have a bunch of light blue (gray in this black and white image), full siblings have light and dark blue (gray and black) and unrelated people have no blue (mostly white). The level of relationship is obvious in each case and there is no need for any statistics.

If you get the result on the upper right with light and dark blue, then the other person is your brother. Period.

This is much more definitive than you might get with a paternity-based DNA test. Then they'd just be able to tell you how likely the man is to be your brother.

For those people who are interested, I want to spend a little time explaining what these results are actually telling you. And how they are related to those colorful rectangles I talked about earlier in the book.

Deeper Science Dive: From Four to One

Let's explore how to go from my four colorful rectangles to 23andMe's single one.

As a reminder, here is an example of what brothers look like with 23andMe:

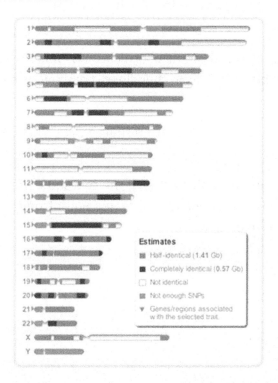

This is a graphical representation of a set of a set of human chromosomes. As you can see there are 23 pinched rectangles (numbered 1-22 plus the X and the Y) but it doesn't look like they are in pairs. It looks like there is just one chromosome instead of two. This is not the case— each rectangle represents a pair of chromosomes (except for the X and the Y).

What they have done is mashed each pair they are comparing into a single rectangle. There are essentially four different chromosomes, two pairs, worth of data in that rectangle.

If they share identical DNA on just one of the pair, it is colored light gray here. This is what they call "half-identical."

If there is identical DNA on both chromosomes in the pair, then it is black. This is, in their lingo, completely identical DNA.

And of course white means no shared DNA. This is their "not identical."

Black can usually only happen if the two share the same two parents. To understand why, let's re-introduce an image I used earlier:

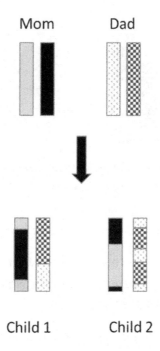

As a refresher, this is a representation of a single pair of chromosomes. Mom has a gray and a black one in her pair and dad has a speckled and a checkered one in his.

They have two children who are each 50% related to their parents. But they obviously inherited different parts of their parents' chromosomes. Their DNA does not look identical.

Again, let's line up the chromosomes from the two kids by putting the ones they got from mom next to each other and the ones they got from dad next to each other:

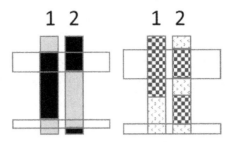

So on the left are the chromosomes each got from mom and the right are the ones from dad. And again, the open rectangular boxes represent where the two of them share DNA across one of the chromosomes in a pair.

A close look shows there are places where the two share parts on both chromosomes. In other words, places where the gray rectangles overlap.

These are the parts that would be in dark blue on the 23andMe representation. Something like this:

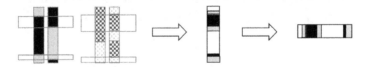

This is a little confusing but I can take you through it.

What I have done is to take the four chromosomes in the right part of the image and combined them into the single rectangle in the middle. (The fourth part of the image is just tipping the rectangle over so it looks similar to the 23andMe one.)

As you can see, the single rectangle now has white, gray and black parts just like what we saw with the 23andMe results. The white parts are where there is no shared DNA (no gray box), the gray is where there is DNA shared on just one pair and the black is where it is shared on both pairs.

Let's look at the top set of rectangles on the left part of the image:

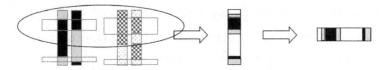

A close look at the circled, empty rectangles shows that this person shares DNA on both sets of chromosomes in the middle part of the box and shares only on dad's chromosomes at the very top and bottom of the box. This is why I made the top and bottom light gray and the middle black:

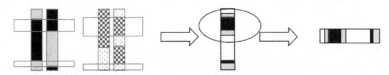

For the DNA shared towards the bottom of the chromosome, you can see that the top part is shared on mom's chromosomes, the middle on both and the bottom on just dad's. That is why there is light gray, black, light gray.

You can also see that this does not distinguish one parent from the other. All we can tell is that the light gray is shared with either mom or dad. To learn more, we'd need to compare it to mom and/or dad's DNA too.

Painting Your Chromosomes

Advice: If you want to get the kinds of images that 23andMe offers with a different testing company, turn to the online site GEDmatch.

There is a way to get these kind of results with companies that don't "paint" your results onto your chromosomes. For this, you need to turn to an online site called GEDmatch.

GEDmatch is a free online service that lets you take your raw DNA data and do lots of fun things with it. I'll quickly go over one of their tools that will be able to tell whether two people are full or half-siblings.

As you'll see, it is very similar to what 23andMe offers. It just takes a little more work.

The first thing you need to do is to download your raw data from the company you used. This will be different for different companies and since how to do it tends to change with each website update, I won't go into the details of how to do it. Check your company's help section to see how to do it.

Once you get the raw data from the two people, you will next sign in to GEDmatch where you will see this in the right hand column:

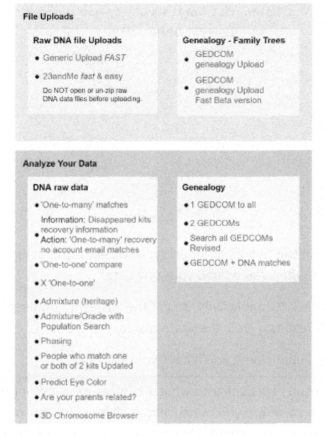

Upload both sets of raw data. Once you do that, click on the one to one compare (under analyze data).

This screen will come up:

GEDmatch.Com DNA one-to-one Comparison Entry Form

This utility allows you to make detailed comparisons of 2 DNA kits. Results may be based on either default thresholds, or thresholds that you provide. Estimates of 'generations' are provided as a relative means of comparison, and should not be taken too literally, especially for more than a couple of generations back.

Kit Number 1:	
Kit Number 2:	
Show graphic bar/numeric positions for each Chromosome?	○ Graphics and Positions ○ Position Only ● Graphic Only
For Full resolution graphic, check 'Full resolution' Window width in pixels	○ Full resolution 1000
SNP count minimum threshold to be considered a matching segment (Leave blank for default value = 500)	

Paste in the two kit numbers (you will get those assigned to you when you upload your raw data).

Next click "Graphic Only" to get the easiest to understand set of results. (Well, these are the easiest for me.)

Click "Submit" and you'll have your results.

The results are similar to what 23andMe gives except that it is green where you match across all four chromosomes, yellow if it just matches across two (from one parent or the other usually) and red where there are no matches across the four chromosomes.

So if you see lots of green blocks of DNA, then most likely you are dealing with full siblings. No green, some yellow and more red means you are dealing with a half sibling.

You can also tell whether you are full or half siblings by the amount of DNA you share. Siblings share around 50% of their DNA while half siblings only share around 25%. The amount shared is usually expressed in something called centimorgans.

Full siblings tend to share around 3500 centimorgans while half siblings share closer to 1750. You can find those numbers at the bottom of the graphic image.

Powerful Tests Can Find Fifth Cousins

Advice: Keep in mind that while the more powerful relationship tests are very good at seeing if two people share a common ancestor, they are not as good at nailing down what that relationship might be, particularly for distant relationships. This means the relationship they assign may not always be the right one.

Tests like those from 23andMe and Ancestry.com are incredibly good at seeing relationships between people. Even distant people like 5th cousins have a pretty good shot at seeing the shared DNA that ultimately came from their great-great-great-great grandparents.

Here is a useful table that breaks out the odds that two relatives will share DNA in one of these tests (along with how many of these relatives you have on average):

Relationship	Likelihood of IBD	Expected number
1st cousins	100%	7.5
2nd cousins	100%	37.5
3rd cousins	98%	187.5
4th cousins	71%	937.5
5th cousins	32%	4,687.5
6th cousins	11%	23,437.5
7th cousins	3.2%	117,187.5

In this table, "likelihood of IBD" is a scientifically precise way to say "chance of being able to see shared DNA." This is the chance that the two people share a chunk of DNA from those distant relatives big enough to see in the test.

These tests are incredibly powerful even up to third cousins. Theoretically, you should see shared DNA in every first and second cousin and 98% of third cousins. It goes down after this but you even have a pretty good chance to see seventh cousins!

This is way more powerful than other relationship tests on the market that are based on the same principles as a paternity test. These could not easily identify shared DNA in even first cousins. As I have said, they have trouble with even siblings!

The table also gives you some idea about how many first, second and so on cousins you might expect to have. This is based on around 2-3 kids per generation and so doesn't fit everyone's situation but it does give you an idea about how the numbers increase. You probably have more than 100,000 seventh cousins for example.

That helps give a feel for how significant a find a seventh cousin is.

What these tests are less precise about is the actual relationship. So sometimes the test will call someone as a third cousin when they may actually be a fifth cousin and so on. The more distant the relationship, the less precise the relationship assignment.

But these tests are still stronger than any others at telling if two people share DNA from distant relatives. They just don't always get the exact relationship right.

Theory vs. Reality. Lots of people do simple math to figure out how related two people are. Using this system, you say you are 50% related to your parents, brothers and sisters, 25% to your grandparents, aunts, uncles, nieces and nephews and so on.

These numbers get a little trickier with cousins. You might say you are 12.5% related to your first cousin, 3.125% to your second cousin, 0.78125% to your third cousin, and so on.

But as I mentioned before, in real life, these are averages. The exact numbers will vary, sometimes a lot.

Earlier I talked about that great site, ISOGG Wiki, where they have compiled a list of the shared DNA between people of known DNA relationships. Let's see what the reality of recombination might do to affect relationship prediction.

Is Johnny your first or second cousins?

Remember, first cousins should share 12.5% of their DNA while second cousins should share 3.125%. Reality is a bit messier.

For example, first cousins have shared as little as 1.2% of their DNA and as much as 22.9%. And when we look at second cousins we see something similar. Here the range is 0.7% to 11.2%.

What this means is that anything between 1.2% and 11.2% could be either a first or second cousin. At the lower end second cousin is much more likely and at the upper end first cousin but either is possible in these ranges.

In fact, for first cousins, the higher number is pretty close to what you might expect for an uncle and a nephew and the lower number is more like third cousin twice removed. This means that in extreme cases, first cousins might look like an aunt and her niece!

The bottom line is that these tests are incredibly powerful at telling if two people share DNA but are less good at identifying the relationship. For example, they will pretty much be able to tell that any first cousins all share DNA and so are related.

But because of those ranges of possible shared DNA, it isn't nearly so easy to parse out the exact relationship. And this matters depending on the question you are asking.

That is where good old fashioned genealogy leg work comes in handy. You have proof that two people share DNA and then you search through various databases to reconstruct the family tree that shows what the relationship actually is.

If, for example, the genealogy all points to first cousin but you share 19% of your DNA, you could still be first cousins. Even if you are tempted to start thinking aunt/nephew or half-siblings.

You need both the paperwork and the DNA results to really nail down the result.

When Your Mom's 4th Cousin Shares No DNA with You

Advice: Ideally try to test older relatives as they may have DNA connections that are lost in the next generation.

DNA tests like the ones I have been talking about can sometimes give what seem to be impossible results. For example, I know of a case where a mom's 4th cousin shared no DNA with her son.

You'd expect that since the son is related to his mom and she is related to this second person that the son would be related to that second person as well. While this is all true on a family tree, it may not be true at the DNA level.

The reason has to do with how much DNA you share with a 4th-6th cousin. And the fact that you only get half of your mom's DNA.

As a start, here is what data looks like from 23andMe for my 4th cousin (note this is their new way of showing this data except that instead of gray, it is purple):

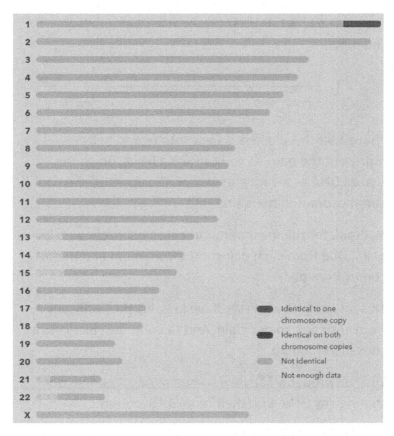

This image represents my 23 pairs of chromosomes.

In this image, each pair (except for the X and the Y) are represented by a long rectangle. So each rectangle is actually a representation of a pair of chromosomes.

The DNA I share with this relative is that dark gray box at the end of chromosome 1. That is all of the shared DNA that this test can see. And although it is hard to tell in a black and white image, it is colored so that the shared DNA is only on one of the chromosome 1's in this pair.

For the next part, I will show the pair of chromosome 1's like this:

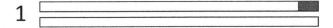

I have split the single rectangle into two to better represent the pair. As you can see, I have shown the shared DNA as a dark gray box on the upper of the two chromosomes in the pair.

As I said, for the most part, our chromosomes come in pairs. One from each pair comes from mom and the other comes from dad.

This means that when we have kids, we pass only one from each pair to our child. And the one the child gets is chosen at random.

You can probably see how someone might end up not sharing any DNA with their mom's 4th cousin.

Imagine that I passed the bottom chromosome down to my son. Now, since this was the only bit of DNA I shared with this 4th cousin, my son won't share any DNA with my 4th cousin.

Here is what his chromosome 1 pair would look like:

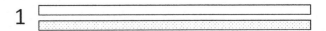

Here my chromosome is the top one and his mom's is the bottom, speckled one. He did not get the shared DNA from me because he got my bottom chromosome and his mom didn't have any of the shared DNA to give. He has no shared chromosome 1 DNA with my 4th cousin.

The 23andMe results comparing my son and my 4th cousin might look like this:

[Chromosome comparison chart showing chromosomes 1-22 and X with no shared segments highlighted]

Now the two look unrelated even though we know they are through me!

This shows one of the limitations of DNA research. You are not going to find all of your relatives because by chance, you may not share enough DNA to see some of them.

It also shows why it can be important to have people tested as far up the family tree as possible. They are more likely to share DNA with people further up and/or out on the tree.

And finally it shows why having multiple people tested in a family can be useful.

For example, imagine that my daughter was tested instead of me. And imagine that she got the top chromosome 1 with the dark gray box.

Now my son would be able to see his relative with her DNA. Of course if my daughter got the same bottom chromosome my son did, then that 4th cousin would still be invisible in this analysis.

Teasing a Little Extra Information Out of your Results

So that is the basics of relationship testing and how it works. But that isn't all you can learn with these kinds of tests.

For example, you can get some idea about whether your parents were closely related or not. And, under the right circumstances, which DNA came from which parent. Both without testing your parents directly.

How to Tell Your Parents are Closely Related

Advice: If you want to find out if your parents are related, you can upload your raw data to an online site like GEDmatch. It is not perfect but it can give you some idea about whether your parents might be related to one another.

It is possible to get some idea about how related your parents are without even testing them. In other words, just by looking at your DNA.

One way you can use your DNA to see if your parents might have been related is through a program like GEDmatch's "Are your parents related?" app:

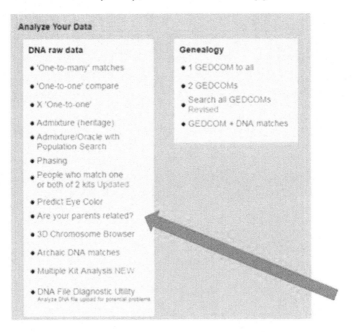

It can look for telltale signs in your raw data from a company like 23andMe or Ancestry.com that your parents are related. And it can give some idea about how related they might be.

The key to how this app works boils down, again, to recombination. As you can probably tell by now, recombination is an important concept for understanding your DNA and your DNA test results!

Let me know if this sounds familiar—our DNA is stored in our cells as 23 pairs of chromosomes. You get half of your DNA from mom and half from dad because you get one chromosome of each pair from mom and one from dad.

Let's look at a person whose parents aren't related. If the parents are unrelated, then the chromosomes in each pair are different from one another.

For example, one pair of dad's chromosomes might look like this:

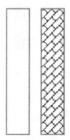

Dad has two different-looking chromosomes because the DNA is slightly different in each. That's because he got one of the pair from his mom and a slightly different one from his dad. (Remember he has 22 other pairs as well.)

Now let's add mom. She also has two different chromosomes.

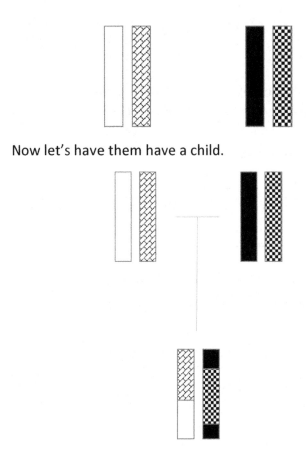

Now let's have them have a child.

In a simple world, one without recombination, you might think that each parent would pass one or the other of the chromosomes in a pair. But as you can see in this image, life is not so simple!

The child does have one copy from her mom, and one from her dad. But before they got handed down to the child, there was a mixing step, recombination, that happened.

When this happens, pieces of the chromosome are swapped between pairs. That's why mom starts out with two different-looking chromosomes and her child ends up with a chromosome that looks like a mix of the two. This swapping happens with dad's chromosomes, too.

When this child has children of her own, the same swapping will happen between their chromosomes.

Imagine that this child has a child with an unrelated partner (in gray):

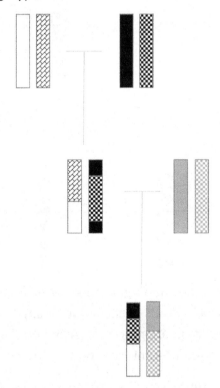

The second-generation child ends up with one chromosome that's a mix of the original mom and dad. The other chromosome is a mix of their other parent's chromosome. And this mixing carries on through the generations.

OK so that is how unrelated parents look. Now let's see what happens when the parents are related.

Let's look at our original set of parents again. Imagine they have two children. Those children both get a mix of dad's chromosomes and a mix of mom's chromosomes in their pair:

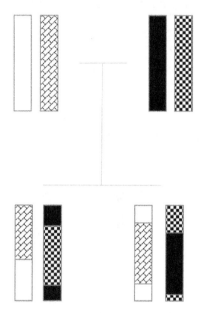

As you can see, they don't get the same exact mix of DNA from each parent. That's because the swapping is different every time recombination occurs. (Again this is why siblings can be so different from one another!)

They do have some pieces in common though. Now let's see what would happen if these two siblings had a child of their own.

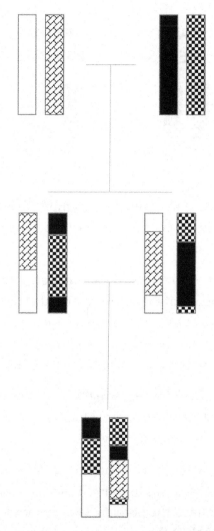

First off, you can see that the swapping happened again. But you can also see a few parts of each chromosome in the pair that look the same.

To make this clearer, I show the child with related parents next to the child with unrelated parents on the next page. DNA that is identical is shown with the empty box.

| RELATED | UNRELATED |
| PARENTS | PARENTS |

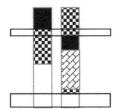 

The areas that are the same in the empty boxes are "runs of homozygosity," which means parts that are the same on both chromosomes.

As you can see, only the child with related parents has any significant amount of this DNA. The child with the unrelated parents has none of these runs of homozygosity. This is because the chromosomes of the child with closely related parents are very similar to begin with.

Scientists can check for these regions in a person's DNA, and count how many there are. Children born to unrelated parents will have relatively few of these areas.

But if there are many runs of homozygosity, this might be a sign that the child was born from closely related parents. And you don't need the parents' DNA to look at this. You only need to look at the child's chromosomes.

Notice the word "might" in the first sentence. This is by no means a foolproof way to figure this out as there can be other reasons for these runs of homozygosity (for example, people of Ashkenazi Jewish descent often have these regions of identical DNA even when parents are not closely related in time).

Mom's DNA or Dad's?

Advice: If you find your half sibling with a DNA test, you can get some idea about which parent you share by looking at the X and Y chromosomes and the mtDNA. 23andMe works well for this.

Sometimes relationship tests help long lost relatives find each other. A common question when half siblings find each other is whether they share a common mom or dad.

It is possible to get some idea about how they might be related.

The first is to look at those famous X and Y chromosomes. If the two people are half brother and sister and share DNA on their X chromosome, then they have a common mom. And if they don't share DNA there, then they may have a common dad.

The second way is for them to look at their mitochondrial DNA (mtDNA) results. This DNA is only passed from mother to child. So if the two of them have different mtDNA, then they probably don't share the same mom.

A third way is if the two of them are half-brothers. In this case, if they share the same Y-DNA, then they may have the same dad. And if they have different Y-DNA, then they almost certainly share the same mom.

Combining these results should give half-siblings a pretty good idea of whether they share a common mom or dad.

If they are half siblings that is.

Keep in mind that a half sibling looks an awful lot like a grandparent and grandchild or uncle/aunt and niece/nephew relationship at the DNA level. And that first cousins can sometimes share enough DNA to get called a half sibling in one of these tests.

So it may be that the two aren't even half siblings.

But if they are, then the X, Y and mtDNA can help them identify whether they share a mom or dad in common. Let's dig into this a little deeper.

X and the Y

As I said, one way you will know if you have a common mom or dad is by looking at the X and Y chromosomes. These are the chromosomes that determine if you will be biologically male or female. Biological males have an X and a Y chromosome and biological females have two X's.

When parents have kids, they each pass one of these chromosomes to their child. So moms always pass an X and dads pass either an X or a Y. If dad passes an X, the child will be a girl biologically and if he passes a Y, the child will be biologically a boy.

Let's first think about a half-brother and half-sister. The half-brother got his Y chromosome from his dad and his X from his mom. His half-sister got one X from her mom and the other from her dad.

Here is a diagram of what this would look like if they have a common dad:

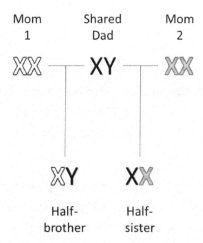

In this diagram, the shared dad has a black X and a black Y chromosome, mom 1 has two white X's and mom 2 has two gray X's.

As you can see, the half-brother has a white X and his half-sister got a gray one. The two of them don't share any X DNA because their common dad gave the half-brother a Y and the half-sister an X. The half-brother's only X came from the unshared mom.

Here is what things look like with a common mom:

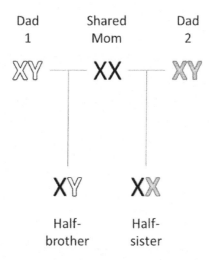

In this diagram, the shared mom has two black X chromosomes, dad 1 has a white X and a white Y chromosome, and dad 2 has a gray X and a gray Y chromosome.

In this case, the mom gives an X chromosome to each child. The half-brother will get his Y chromosome from his dad (Dad 1), the half–sister will get her other X chromosome from her dad (Dad 2). So, in this case, the half-siblings will share X chromosome DNA. They both have black X's.

So if the two of them share X DNA, they almost certainly share the same mom. But it turns out that biology is a bit more complicated on the other side.

The image here is not quite right in that mom's two X's aren't really identical. She should not have two black X's. So here I have drawn her DNA as having one black and one very light gray X:

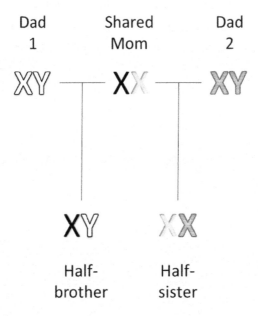

Now here you can see that these two half-siblings actually don't share any DNA on their X's because the half-brother got the black X and the half-sister got the very light gray one. From this diagram it would seem that this should happen 50% of the time.

But it doesn't. It is actually very uncommon for half-siblings who share a common mom to have no shared DNA on their X chromosome. The reason is that old favorite, recombination.

Recombination involves DNA swapping which means that DNA is swapped between the two X's in women. Each half-sibling would get an X from mom that has some black and some very light gray. The odds that there is no overlap is actually very small.

But it is real. Which is why you want to look at mitochondrial DNA (mtDNA) too. (Keep in mind that Ancestry.com does not offer this as part of their DNA test.)

Mitochondrial DNA

Mitochondrial DNA (mtDNA) is passed only from mother to child. All of dad's mtDNA is destroyed right after fertilization.

As you might recall, this type of DNA is found in ancient structures in our cells, called mitochondria, also referred to as the "powerhouse of the cell." It has its own DNA that helps it to function, and all humans ONLY get mitochondria from moms!

So, half-siblings can compare their mtDNA to verify that they share a common maternal ancestor. If they share a mom, they should have the same mtDNA. And if they don't, then their mtDNA would be different.

Here is what these sorts of results look like on 23andMe:

You can find this in the ancestry section under haplogroups. This person (me) has the maternal haplogroup 2a. If I had a half-sister through a common mother, then she would be H2a as well.

It is important to keep in mind that lots of people share this particular haplogroup including pretty distantly related people. Many thousands or even millions of people have this exact same haplogroup. So if two people share the same maternal haplogroup, they are not necessarily closely related.

But the test can be used to show that two men do NOT share the same mother if they have different maternal haplogroups.

Another important thing to keep in mind is that just because a test says you are half siblings and you share the same maternal haplogroup, that does not mean you are.

Imagine you have a first cousin through your mother's sister like this:

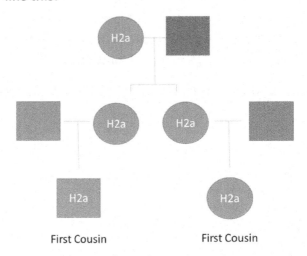

Ok, this family tree looks different than others I've drawn. In this family tree, circles are females and squares are males. We are following the mtDNA in this tree focusing on grandma's H2a. Unlabeled squares mean other maternal haplogroups (different mtDNA).

Grandma is H2a and grandpa's does not matter in this case.

They have two daughters who are also H2a. If they had had a son, he'd be H2a as well. All their kids have the H2a mtDNA haplogroup from mom.

Each of these daughters marry a man with a different maternal haplogroup (unlabeled squares). One has a son and the other a daughter. Both children are H2a like their moms—we have two first cousins with the same mtDNA.

Now think back to the DNA test they took. The test called them as half-siblings and, since they have the same mtDNA, they conclude they share a common mom. But we know this isn't true (because we made up the family tree).

The test miscalled them as half-sibling because they happen to share a lot more DNA than first cousins are predicted to. So they are first cousins with sisters as moms, not half-siblings with a shared mom.

It is important to keep these sorts of things in mind as you go over your test results.

Half-Brothers and the Y

In the special case of two half-brothers, you can look at the Y-DNA to see if they share a common dad. (Again, this is not offered by Ancestry.com.)

As I have said before, it is usually only men that have Y chromosomes which means that dads pass this DNA only to their sons. I have also mentioned that this DNA (for the purposes of our discussion) does not have a partner with which it can recombine meaning that the Y changes very little when it passes from father to son.

So here is what it looks like when two half-brothers share the same dad:

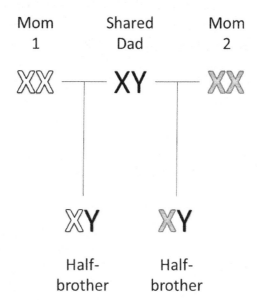

They share the exact same Y. Here is what the Y-result (paternal haplogroup) looks like with 23andMe:

You can find this in the ancestry section under haplogroups. This person (me) has the paternal haplogroup R-L48. If I had a half-brother through a common dad, then he would be R-L48 as well.

It is important to keep in mind that lots of men share this particular haplogroup including pretty distantly related men. Many thousands or even millions of men have this exact same haplogroup. So if two men share the same paternal haplogroup, they are not necessarily closely related.

But the test can be used to show that two men do NOT share the same father if they have different paternal haplogroups.

For completeness sake, here is what it would look like with two half-brothers who share a common mom:

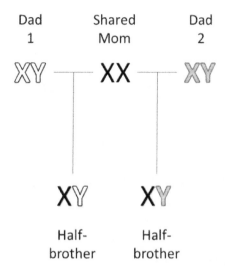

Here they have different Y's meaning they would have different paternal haplogroups. So if you get this result and know you are half siblings, odds are you share a common mother instead of a father.

Again, this only applies if they are actually half-brothers. Remember back to our mtDNA example where half siblings were actually first cousins? Something similar could happen with Y-DNA too.

Imagine this situation:

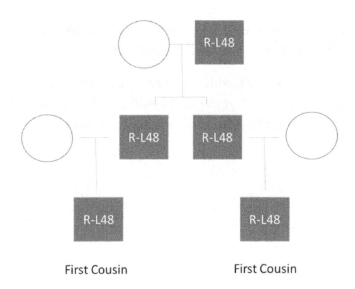

First Cousin   First Cousin

Again, squares are males and circles are females. Here we are following some Y-DNA, the haplogroup R-L48. Since the women do not have a Y chromosome, they do not have a paternal haplogroup in their DNA and so I have left them blank. (If they wanted to know their family's paternal haplogroup, they would need to test their dad or a brother.)

As you can see, the R-L48 passes from grandfather to father(s) to son(s). So we have two male first cousins with the same Y-DNA.

Just like with the mtDNA, think back to the DNA test they took. The test called them as half-siblings and, since they have the same Y-DNA, they conclude they share a common dad. But we know this isn't true (because again we made up the family tree).

The test miscalled them as half-siblings because they happen to share a lot more DNA than first cousins are predicted to. So they are first cousins with brothers as dads, not half-siblings with a shared dad.

Hopefully these two examples have shown the dangers of diving too deep into your results without having a deeper understanding of DNA and how it is passed on.

Powerful Relationship Tests Can Be Wrong

Here is a really interesting example where all sorts of tests went wrong for paternity. It wasn't because the testing companies failed. Instead, it was because of a rare biological situation called chimerism.

A man and his wife went to a fertility clinic and were worried that the wrong embryo may have been implanted. They both have an A blood type and the child had an AB blood type which, as you can see in Appendix A, is a paternity red flag.

Now while it is a red flag, it does not rule out paternity completely. There are rare exceptions to this rule.

But most of these rare exceptions did not seem to be to blame as a couple of DNA paternity tests also showed he was not the father. The chances for paternity were zero.

The paternity test turned out to be right. And to be wrong at the same time.

Remember, one problem with standard paternity tests is that they are not powerful enough to easily see a relationship like uncle/nephew. What I haven't talked about yet is another drawback of these tests.

See, these paternity tests tend to focus pretty exactly on the question they are being asked. The question here was, "Is this man the father?" The answer to this was no.

What was hidden though, because of the weakness of the test and the fact that it wasn't the question getting asked, was the actual relationship between the man and the boy.

To see this, the couple turned to one of the more powerful tests offered by 23andMe.

Remember, here is what a test like this would look like if there was truly no relationship between the man and the child:

It is all grayed out with no dark gray/black which indicates no shared DNA between the two of them. (The real test has purple instead of dark gray/black.)

And here is what it would have looked like if they were related as father and son at the DNA level:

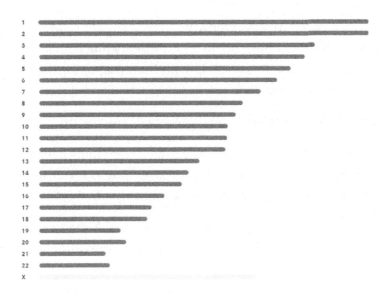

Here all the chromosomes are dark gray/black because the two of them share DNA all across one set of chromosomes.

Here are the results they actually got:

As you can see, this is somewhere between not related and related as father/son. In fact, this very much looks like an uncle/nephew relationship with the uncle being the brother of the real dad.

So this man's brother is the dad? Again, yes and no.

It turned out that the dad was actually something called a chimera. What that means is he started out as fraternal twin boys who fused when they were only a few cells big.

The resulting person, the father in this case, is made up of a set of cells from one twin and a set from the other. He has two sets of DNA!

I show how one way this might have happened on the next page:

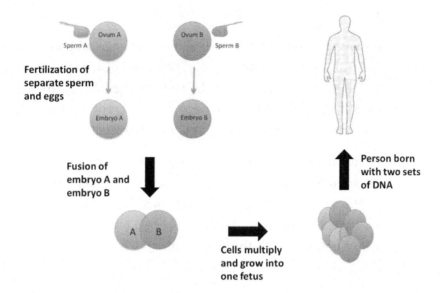

In fraternal twins, mom releases two eggs, each of which is fertilized by a different sperm. In this case, the eggs are "Ovum A" and "Ovum B" and the sperms are "A" and "B."

What usually happens is that embryo A and embryo B grow into two unique people. They are fraternal twins.

What happens with chimeras is that the embryos fuse at a very early stage. Now embryo AB is made up of two sets of cells, one from embryo A and one from embryo B.

So our dad has cells that have the DNA from one brother and cells with DNA from the other brother. He really is father and uncle to his children!

The DNA from his cheek cells came from one brother and the DNA in the sperm that resulted in his son came from the second brother. This is why, even though the dad is a single person, he looks like his son's uncle.

This was shown to be the case when a testing company looked at the DNA in his sperm. They saw both sets of DNA, one from each brother. By chance, the DNA of the sperm that fertilized his son's egg was different than the DNA of the cells that were tested, the cheek cells.

And to add another twist, it turns out his other son happened to be conceived with a sperm that did match the DNA in his cheek cells. His two sons share less DNA than two brothers typically do! It is like a man and his brother had kids with the same woman.

Artificial Chimeras

Advice: If you have had a bone marrow or any other transplant, let your testing company know. It could affect your results even with a cheek swab.

While the chimeras I talked about in the previous section are thought to be rare, other, artificial ones, are not.

One way to be a chimera is if you get a bone marrow transplant. In this case, your blood has the DNA of the donor while the rest of your cells have your DNA.

You can imagine what might happen if you commit a crime and leave DNA at the scene of a crime. Now it will look like the donor committed the crime not you.

In fact, back in 2004, Abirami Chidambaram of the Alaska State Scientific Crime Detection Laboratory in Anchorage gave details about one such case. The case involved a serious sexual assault.

The detectives found that a semen sample from the crime scene matched the DNA found in the blood of someone in the criminal database. They had caught their criminal!

Except that the man had an ironclad alibi—he was in jail when the crime was committed. So he couldn't have done it.

A little digging showed that the man in jail had received a bone marrow transplant from his brother. So he and his brother had the same DNA in their blood.

So the DNA in the semen actually matched his *brother's* DNA, the donor. When they tested different cells, they quickly saw that the man in jail did not commit the crime. The culprit was his brother.

It seems like this might only happen with blood tests but it turns out it can happen with cheek cells too. There can sometimes be enough cross contamination between blood and cheek to cause mix-ups like this.

And of course this sort of thing could happen with a paternity test as well. Which is why good paternity testing companies now ask people if they have had a bone marrow transplant before they are tested.

If you are involved in a paternity situation, definitely make sure people know if you have had a bone marrow transplant or not.

Part 3: Ancestry: Where'd I Come From?

A very popular, recreational use of DNA testing right now is ancestry testing. You can tell a lot about your family's history by looking at your DNA.

But you can't tell everything. And the results you get back can be very confusing.

For example, it is perfectly reasonable from a scientific perspective for siblings to have different DNA ancestry results. This is even though they share the exact same parents.

To get the most out of these tests, you need to have a deeper understanding of what they look at and how they work. I try to do that in this last part of the book.

Some Ancestral Testing Advice

1) Keep in mind that these tests are not precise. Just because they give you an exact percentage that does not mean the percentage is written in stone. There is still a lot of wiggle room in these results. So if you get back 50% Scandinavian, interpret that as a lot of Scandinavian. Not exactly 50%.

2) Because of how DNA is passed down, you and your sibling can have different ancestry results. This does not mean your sibling is not your brother or sister. This is just a consequence of how DNA works.

3) Mitochondrial and Y DNA testing are fun but not always that informative because, again, of how they are passed down.

4) Some companies (like 23andMe) offer you a chance to see how Neanderthal your DNA is.

What Autosomal DNA Ancestry Tests Look For

Advice: These results are powerful but don't get too hung up on the percentages. Instead, think of 30% "East European" as a good bit of "East European."

DNA tests like Ancestry.com look at the parts of your autosomal DNA that are more likely to be the same between people from the same parts of the world. Autosomal is the scientific term for the chromosomes and excludes the mitochondrial DNA (mtDNA) and DNA from the Y-chromosome (Y-DNA). I'll talk about those DNA later.

What I want to do first is talk a little about why people from the same part of the world share some of the same DNA.

Ancestry tests work by tracking differences in DNA between people. They can do this because our DNA gradually changes over time.

For example, when cells copy their DNA, they don't always do a perfect job. Mistakes called "mutations" happen and not everyone has the same mutations in their DNA.

In fact, people from the same part of the world tend to share some DNA mistakes that people from different parts of the world don't have. This is because thousands of years ago, humans didn't move around as much as we do today. What happened to DNA in one part of the world stayed in that part of the world.

This means different groups built up different sets of mutations over time.

Picture your DNA as a long book. In the old days before computers and modern technology, books were copied by hand. Monks in monasteries carefully transcribed new copies of books, but sometimes they made mistakes.

Imagine that a monk in a monastery in Poland is copying a book and makes a mistake. These monks then lose the original and use the copied book to make new copies. Now all the copies of that book in the Polish monastery will have that new mistake.

The same thing happens to your DNA. When there is a "mistake" or mutation in the DNA that gets passed down to a child, then that child will have that mutation. As will all of his or her kids.

Here's the important part, though — all kinds of different mistakes can happen.

For example, getting back to our Polish monk, if he makes a mistake, that mistake will be in all of the Polish books from then on. But the Polish monk will probably make different mistakes compared to an English monk.

And a historian would be able to trace a book back to the monastery it came from by looking at the mistakes. If the book has the Polish mistake, it probably came from Poland. If it has the English mistake, it came from England.

It's the same with DNA. People who share ancestors from the same part of the world are more likely to have the same mutations in their DNA.

And just like historians can trace where a book is from based on its mistakes, scientists can trace where you are from by looking at mutations. If you are Eastern European, then you have the same "mistakes" in your DNA as other Eastern Europeans.

So that is why they can trace your ancestry using DNA. Here are my 23andMe results:

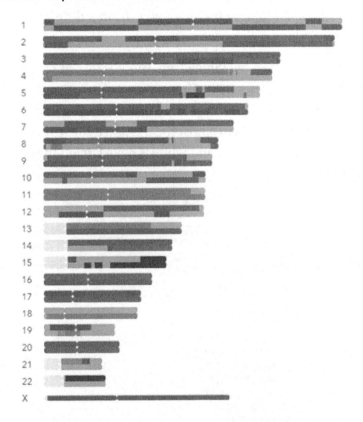

These results are nice and visual and have been painted onto my DNA indicating what part of the world that DNA is likely to be from. (The real results look better as they are in bright colors.)

Remember, we have 23 pairs of chromosomes and we get one from each pair from mom and one from dad. My DNA is arranged into 22 pairs (labelled 1-22) plus my X chromosome. They do not look at my Y for this particular analysis.

Each shading (color in the original) is from a different part of the world.

As expected from my family tree, there is lots of European scattered across my DNA. There is a smidge of South Asian at the tip of chromosome 4 and a bit of sub-Saharan African at the tip of chromosome 11. Since these last two are hard to see in black and white, I have circled them below:

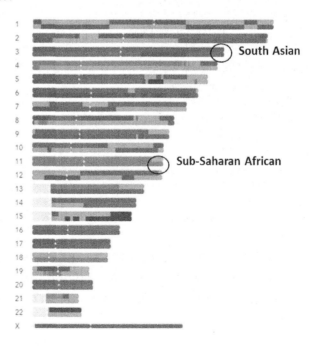

It is really important to not get hung up on the exact amounts of ancestry that you see in the image. Figuring out where one ancestry starts and another ends is very difficult and so not always accurate.

It is also important to take that South Asian and sub-Saharan African with a pinch of salt. These are such tiny bits of ancestry that they could simply be the algorithm misreading my DNA. I could easily have no African in my DNA despite these results. (And in fact, as you'll see later, my Ancestry.com results see no African in my DNA.)

You can also get percentages of various ethnicities, essentially by summing this up across the whole genome. For example, here is my European heritage according to this 23andMe report:

European	99.8%
● Northwestern European	66.0%
British & Irish	25.0%
French & German	10.2%
Scandinavian	2.3%
Broadly Northwestern European	28.5%
● Eastern European	14.2%
● Southern European	4.8%
Balkan	2.1%
Iberian	0.3%
Broadly Southern European	2.5%
● Ashkenazi Jewish	< 0.1%
● Broadly European	14.7%

A key point again is not to focus too much on these exact percentages. They really are just a ballpark figure. In other words, there is a lot of wiggle room in these results.

A different testing company, Ancestry.com, does a much better job at letting people know how much wiggle room there is. If the customer knows how to find the information that is.

Let's use my results as an example. Here is what they look like from Ancestry.com:

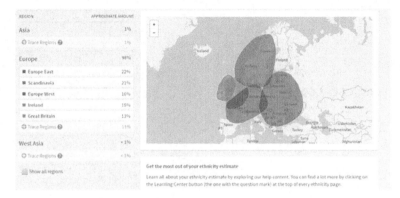

Again the results are much prettier when they are in color!

Notice these are not painted onto an image of chromosomes. Instead you get the total percentage from each group and a map of where the DNA that represents each ethnicity is typically found.

As the numbers for both companies can be hard to see, I have blown them up and displayed them one on top of the other on the next page:

Ancestry.com

REGION	APPROXIMATE AMOUNT
Asia	1%
⊕ Trace Regions ❓	1%
Europe	98%
■ Europe East	22%
■ Scandinavia	21%
■ Europe West	16%
■ Ireland	15%
■ Great Britain	13%
⊕ Trace Regions ❓	11%
West Asia	<1%
⊕ Trace Regions ❓	<1%

23andMe

European		**99.8%**
●	Northwestern European	66.0%
	British & Irish	25.0%
	French & German	10.2%
	Scandinavian	2.3%
	Broadly Northwestern European	26.5%
●	Eastern European	14.2%
●	Southern European	4.8%
	Balkan	2.1%
	Iberian	0.3%
	Broadly Southern European	2.5%
●	Ashkenazi Jewish	<0.1%
●	Broadly European	14.7%

The first thing to notice is that it isn't necessarily that easy to compare the two results. They have diced and sliced my European heritage in different ways.

Where there are the same designations, you can see that the percentages are not the same between the two companies. This is not surprising as they both use different databases and slightly different algorithms.

So for example, I am 22% East European with Ancestry.com and 14.2% with 23andMe. These numbers give some idea of the reasonable ranges involved.

This means no one should worry too much about even 10 percent differences. This becomes more obvious with the results from Ancestry.com when I click on a particular ancestry and the following image comes up:

This is a little hard to see so I will break it up into two separate images:

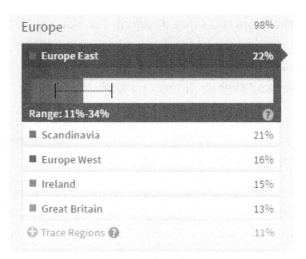

Remember my original report said 22% East European but the actual range is 11-34%. It is most likely around 20% but could be only 11% or up to 34%. This range is consistent with the results I got from 23andMe.

Here is part of the right hand side of the results:

Note on the map that even though "Europe East" is centered on Eastern Europe, the outer fringes, the lighter gray, are in Sweden and Switzerland! My Eastern European DNA might be Swiss.

So if you know where to look, Ancestry.com does a very good job of letting you know the wiggle room in what they present. Both in terms of percentage and the area their designation covers.

All of this is important to keep in mind when you interpret your results. Percentages are general, not rock-solid numbers and Great Britain can go all the way to the Czech Republic.

This is not to say that these companies are doing a poor job. They are doing an excellent job with a technically challenging task in a rapidly evolving field.

OK, so that is the wiggle room in the data. But some real biology can also make an ancestry result confusing. Like when your sister is Swedish and you aren't.

In the next section I'll talk about how that might happen.

Same Parents, Different Ancestry

It seems like brothers and sisters should have the same ancestry background. After all, they both got half their DNA from mom and half from dad.

But because of how DNA is passed on, it is possible for two siblings to have some big differences in their ancestry at the DNA level. Culturally they may each say they are "1/8th Cherokee" but at the DNA level, one may have no Cherokee DNA at all.

To help make this clearer, I am going to turn to another successful analogy—DNA as a bunch of colored beads.

As we have already discussed, DNA isn't passed down from generation to generation in a single block. Not every child gets the same 50% of mom's DNA and 50% of dad's DNA. (Unless of course they are identical twins.)

This has consequences in terms of how much DNA siblings share. And even more significantly, what DNA they share.

One way to think about this is to imagine DNA as a bunch of colored beads. Since we are interested in ancestry here, we will say that different colors mean different ancestries.

Imagine that a man from Japan marries a woman from Europe. Her DNA happens to be 100% European and his 100% Asian.

Let's say that the European beads are black and the Asian beads are white. Here is what this might look like:

Mom ●●●●●●
●●●●●●
●●●●●●
●●●●●●
●●●●●●
●●●●●●
●●●●●●
●●●●●●

Dad ○○○○○○
○○○○○○
○○○○○○
○○○○○○
○○○○○○
○○○○○○
○○○○○○
○○○○○○

When these two parents have a child, that child will get a random half of mom's beads and a random half of dad's. The child might look like this:

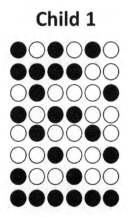

Child 1

This child is 50% European and 50% Asian. Here is what this child's sibling might look like:

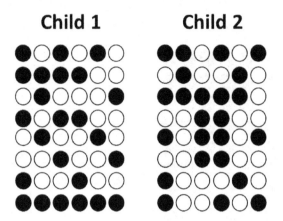

Going across the top row, both got European ancestry from mom for the first bead, child 1 got Japanese from dad and child 2 got European from mom for the second bead, and so on.

In this case you can see that each child actually shares the same ancestry even though they got some different DNA from each parent. They are each 50% European and 50% Asian.

This is the result if the parents happen to be 100% of an ethnicity. It is a different story if the parents have a much more mixed background.

Let's now imagine a couple of more complex situations. First we will add in some gray beads from sub-Saharan Africa. Imagine these parents:

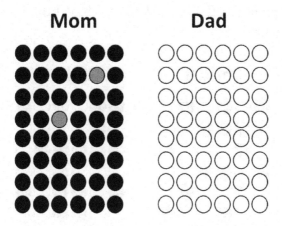

The difference here is that mom has a small bit of African ancestry in her family tree. This is actually pretty common in the U.S.

Otherwise everything else is the same. Black is European, white is Asian.

Imagine this is their first child:

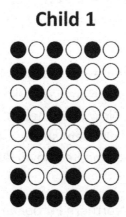

As you can see, by chance this child didn't happen to get any gray beads. A DNA test would say the child has 0% African DNA even though mom has around 4%.

Now imagine a second child happens to get this arrangement:

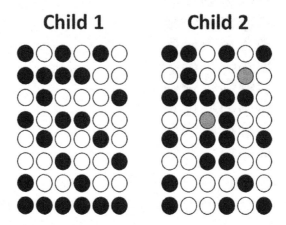

As you can see, by chance the second child inherited both gray beads from mom. In an ancestry DNA test, child 1 would be 0% African while child 2 would be around 4%. Even though they have the same parents.

It all has to do with which DNA you happen to get from each parent. It might have ended up that each child got one gray bead but it doesn't have to be that way.

Now imagine the newer tests that split European, Asian, Pacific Islander, African, Native American and so on into many subcategories. Now even the first mom's all black becomes a variety of different shades of black, white, and gray. And these can get passed down differently leading to different percentages.

Mom might be 23% Northern European, 46% Eastern European, and 31% Southern European. And these can be further subdivided. Maybe the Northern European is really 15% British, 5% Scandinavian and 3% German/French.

Here is what mom looks like now:

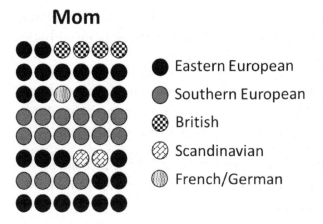

Now she has children with her Japanese husband (who remember has all white beads). Here is what the children's DNA might look like:

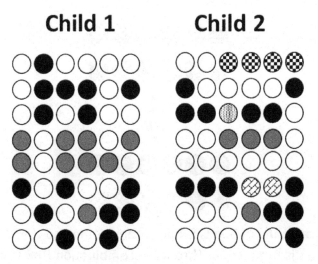

Now it is getting very confusing. By chance, child 1 has no British, French/German or Scandinavian DNA while child 2 does. Even with the same parents!

Now split the Southern and Eastern European DNA into subgroups and see what happens. Then start splitting dad's DNA further to add more to the mix. Do all of this and you can see how siblings might wind up with very different ancestry results indeed.

Of course, real DNA isn't beads. But you can think of it as sort of like beads on a string. Each string of beads is a chromosome, a long stretch of connected DNA.

As I keep saying, people typically have 23 pairs of chromosome. One from each pair comes from mom and one from dad.

Before mom passes down her chromosome to her child, her two chromosomes in a pair swap some DNA. (Yes, this is our old friend recombination coming back into the picture.) Something like this with beads:

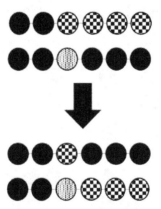

The top string of beads is one chromosome and the bottom is the second. Through recombination (the black arrow), three of the checkered beads at the end of the top chromosome swapped with the three black ones in the bottom chromosome. One of these new chromosomes, a string of beads all in a row, is then passed on to the child.

This process happens with each pair of chromosomes. And happens in new combinations with each egg (or sperm for dad). Infinite combinations are possible! Which is why two siblings can end up with different ancestries with the same parent.

And now back to real DNA!

How You Can Be 55% Italian When One Parent has No Italian DNA

I recently saw a case where a man had a genetic test that claimed 55% Italian. This was confusing to him because when he had his parents tested, one had no Italian DNA (the other did have some Italian DNA).

If the 55% were a rock solid number then this would be very confusing as both parents would have to have some Italian in their DNA. After all, you only get 50% of your DNA from one parent! At least 5% of his Italian heritage would have had to come from the other parent.

But in the real world of DNA ancestry testing, that percentage is a bit more wishy washy. I think of them as saying that a sizeable part of his heritage looks Italian in the DNA test. Could be lower, could be higher.

If the number is more like 40%, then it could easily have come from just one parent. Even if that parent was only 40% Italian too!

It is also important to keep in mind that "Italian" might be broader than the current borders of Italy. It might include France or Switzerland or even a country further afield.

Let's use my examples from Ancestry.com and 23andMe to see what we can squeeze out of each company's results.

Ancestry.com

As I showed previously, Ancestry.com has a very useful resource that gives you the likely range of your percent ancestry. In other words, it lets you know how reliable that 55% is and what its upper and lower limits are.

If I click on my "Scandinavia" ancestry in my results, the following image comes up:

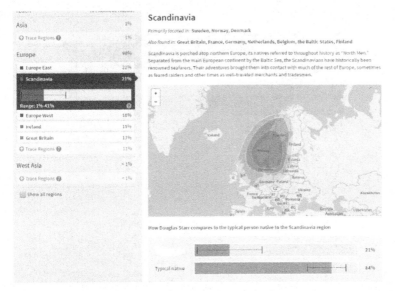

This is a little hard to see so I will break it up into two separate images:

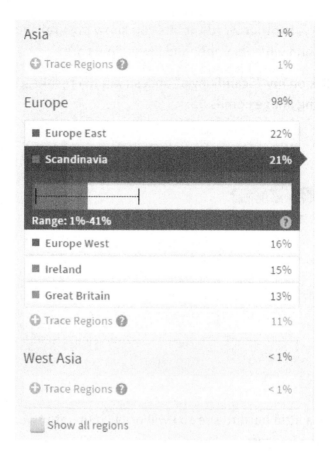

My original report said 21% Scandinavian but the actual range is 1-41%. It is most likely around 20% but could be way more or way less.

If the man clicked on his 55% Italian and got a 30-70% range, then at the lower end, all of the Italian could have come from one parent. The algorithm is saying that 55% is the most likely number but reality could be closer to the 30% number.

Here is the right hand side of my results:

Scandinavia

Primarily located in: Sweden, Norway, Denmark

Also found in: Great Britain, France, Germany, Netherlands, Belgium, the Baltic States, Finland

Scandinavia is perched atop northern Europe, its natives referred to throughout history as "North Men." Separated from the main European continent by the Baltic Sea, the Scandinavians have historically been renowned seafarers. Their adventures brought them into contact with much of the rest of Europe, sometimes as feared raiders and other times as well-traveled merchants and tradesmen.

How Douglas Starr compares to the typical person native to the Scandinavia region

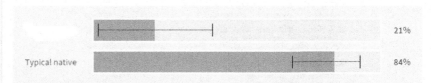

Note on the map that even though "Scandinavia" is centered on Scandinavia, the outer fringes, the lighter shades of gray, are in the UK and Poland! My Scandinavian DNA might be Polish.

So something similar could be happening in our Italian case. What is described by the program as Italian may be described as Greek a different time.

If you know where to look, Ancestry.com does a very good job of letting you know the wiggle room in what they present. While 23andMe doesn't give this sort of information directly, it provides a way to directly see if you got a certain ethnicity from both parents if your DNA lines up just right.

23andMe

As I said, if you do a 23andMe test, then you won't get the range of possible percentages.

What 23andMe does provide is a tool that can help you figure out if an ancestry likely came from one or both parents. They do this by "painting" your ancestry results onto your chromosomes.

Here is a picture from 23andMe of my results:

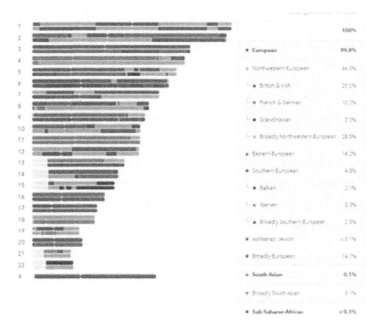

What they have done is shown my ancestry on my DNA. What is represented is 22 chromosome pairs labeled 1-22 and my single X. (They do not show the Y that goes with my X in this image.)

Notice there are two chromosomes for each number. One of each pair comes from mom and one from dad.

There is a lot going on in this image but here is what it looks like if I hover the cursor over just Scandinavian:

On pair 15, you can see that both chromosomes in the pair have Scandinavian segments that overlap. I have circled this region in the image on the next page:

When an ethnicity overlaps like this it usually means that it came from both parents. Remember, the top chromosome came from one parent and the bottom from the other.

The key is the overlap. The fact that Scandinavian is on the top chromosome in pair 22 and the bottom one in pair 5 doesn't mean each came from a different parent.

In my case in particular it is hard to assign which chromosome in each pair comes from mom and which comes from dad. I could do this if I had either of my parent's DNA but I do not.

Without my parent's DNA to compare with, there is no easy way to tell which chromosome came from which parent. Unless of course the parents have two very different ethnicities...something like sub-Saharan Africa and Japanese. (And even then it can be wrong.)

Our 55% Italian man could do something similar with his Italian to see if he got it from both sides.

OK that was fun! Now let's circle back to the idea of a parent and a child having identical amounts of DNA from a certain ancestry. And let's look at chromosomes.

Parent and Child with Same Percentage Ancestry

Here I'll try to explain a situation where one parent has no Italian, the second parent has 40% Italian and the child has 40% Italian.

To understand this, we need to revisit (again!) how DNA is stored in our cells. And how it is passed on.

DNA is stored in chromosomes and most people have 46 of these. They come in nearly identical pairs. One from each pair comes from mom and one from dad.

Imagine these are the 23 pairs:

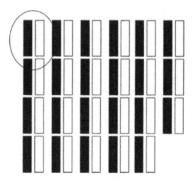

I have highlighted the first pair by circling it.

Let's say the black ones came from dad and the white ones from mom like this:

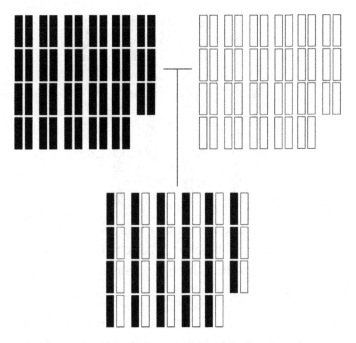

This child got one black from each of dad's chromosome pairs and one white from mom. Which of each pair they get is chosen at random.

Now I am going to overlay some ancestry onto dad's chromosomes. To keep things simple let's say he is 40% Italian and 60% sub-Saharan African.

I'll make the Italian black and the sub-Saharan African gray. I'll also make it so whole chromosomes are either Italian or sub-Saharan African. It doesn't have to work this way, it'll just make things easier.

Here is dad now:

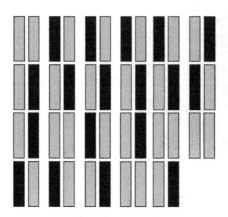

Sometimes he has one from a pair that looks Italian and the other that looks sub-Saharan African (one black and one gray). And sometimes he has a pair that both look sub-Saharan African (both gray).

When he has a child, he passes one chromosome from each pair down to that child. Again the one passed down is chosen randomly. In our image, it could be the one on the left or the one on the right in each pair.

Let's say that by chance they were passed down like in the image on the next page:

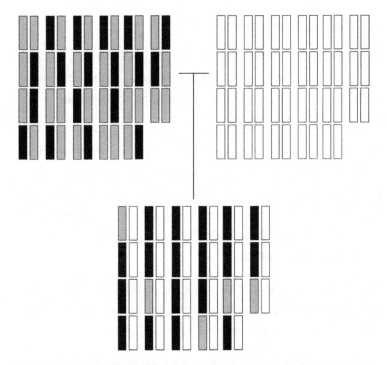

By chance, this child happened to get all of dad's "Italian" DNA (the black chromosomes). The child is 40% Italian just like dad.

Now as you know by now, this is a simplification of DNA passing because we've left off nature's complication—recombination. But the basic idea behind it is sound. This is how parent and child could both be 40% Italian even if the other parent has no Italian DNA.

My Sister is African but I'm Not!

I know of a case when a brother had 0.5% African DNA but his sister didn't have any. So in this case, if only she had been tested, the family would not know about their possible African heritage.

Assuming the 0.5% is real (which it may not be given how small a number it is), it turns out that this is not a surprising result for some of the reasons we already talked about. There is no reason to expect that both siblings should have 0.5% African DNA even though they came from the same parents.

Again this goes against the way most of us think of these things. For example, in my family, we are always quoting what percentage Native American we are.

We had a distant relative who was supposedly from the Cherokee tribe. Because of that, we say that my dad is 1/8th Native American and I am 1/16th. While this may be true from a genealogical standpoint, it doesn't have to be true from a DNA one.

Because of how DNA is passed on, I may have no Cherokee DNA left in my DNA. This is even though my relative really was Cherokee. And even though we really do get half our DNA from our moms and half from our dads.

What we need to know to understand results like these is that the half the DNA we get from each parent is chosen at random. So, for example, it may be that my grandpa just happened to pass no Cherokee DNA to my dad. A DNA test would say he had no Native American in his heritage even though our family tree clearly does.

This sort of thing becomes more and more likely as the amount of DNA gets smaller and smaller. Once we are dealing with 0.5% of something, we are really talking about one or maybe two tiny bits of DNA.

In fact, it is so small each child might have just a 50% chance of inheriting any of it from their parent(s). And if a child happened to get half or even less of that 0.5%, the test might not even see the DNA any more.

So this kind of result is perfectly reasonable. One sibling happened to get a bit of African DNA from one or both of parents and his sister didn't.

As I keep saying, our DNA is stored in chromosomes. Most people have 23 pairs of chromosomes for a total of 46.

One chromosome from each pair comes from mom and one from dad. This is where the 50% of your DNA from mom and 50% of your DNA from dad comes from.

Here again is what my chromosomes look like from an ancestry point of view (my 23andMe results):

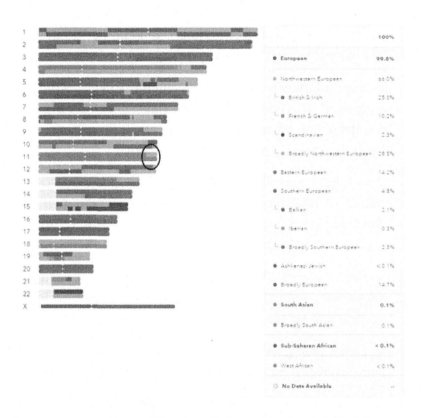

As you can see I have 22 pairs of chromosomes labeled 1-22 and then a lone X chromosome. In reality, I have a Y to go with my X but this isn't shown in this ancestry test. If I were biologically female, I'd have two X's.

One of each pair of chromosomes came from my mom and one from my dad. Because it is hard to see in black and white, I have circled the tip of my chromosome 11 which represents my 0.1% African.

Let's focus on the chromosome 11 pair in the image on the next page:

11

Here I have redrawn the chromosomes emphasizing the African DNA at the tip of the bottom chromosome in black.

I also gave one chromosome a dotted outline and the other a solid outline so we can keep track of it better.

Now looking at me you'd never guess there was any African American in my family tree. But that is what that black at the tip of one of my chromosome 11's suggests (although the number is so small it may not be real—it could just be noise. The idea of noise is supported by the fact that my Ancestry.com data shows no sub-Saharan African.)

Somewhere on my mom's or dad's side of the family, there may have been an African ancestor. It probably comes from my dad's side for family historical reasons.

OK, now let's imagine what might happen with this chromosome pair and my kids. In a simple world, each child would have a 50% chance of getting the top chromosome and a 50% chance of getting the bottom one.

This would mean that each child would have a 50% chance of inheriting a bit of African DNA from me and, of course, a 50% chance of getting no African DNA.

This would be enough to explain the situation where one sibling has African DNA and the other doesn't. One of their parents had one chromosome in a pair that had some African DNA.

One sibling happened to get this chromosome and the other got the other one. Simple.

Except of course, as we've seen before, nothing in biology is ever that simple! To get at what actually happens, we need to go a little deeper into how a parent's DNA is distributed into sperm and eggs.

This explanation won't change the basics of what we've already talked about. But it can show how other results are possible. Like how these two siblings might have each ended up with 0.25% African DNA.

Here we need to come back yet again to recombination.

When we pass our DNA down, we don't actually pass down one of our two chromosomes in each pair. Instead we pass down a chromosome that is a mix of the two in the pair. This DNA swapping is called recombination.

Let's focus on my pair of chromosome 11's. Here is one way the DNA might have swapped:

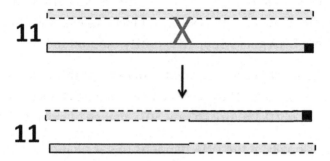

The X between the two in the pair is a common way to show where the DNA was swapped between the two chromosomes.

As you can see, here it doesn't matter that the DNA recombined. Each child still has a 50% chance of getting all of my African DNA and a 50% chance of getting none.

But imagine the DNA swapped like his instead:

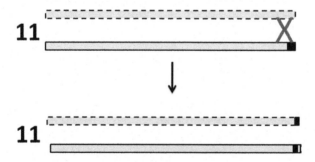

Now each child will get a bit of African DNA. They will be down to 0.05% but it will be there (although it might be a small enough bit that the test can't see it).

A recombination that happens right there, right in the middle of my African DNA, would not happen very often. Where DNA gets swapped is pretty random too.

So the most likely outcome for my kids at this point is an all or none one. They will each have a 50% chance for getting my 0.1% African or a 50% chance of getting none.

Occasionally, if the DNA happens to get swapped in the exact right place, a child has a 100% chance of getting a bit of African DNA. But given how little African DNA I have, the odds are very low for this outcome. Still it could happen.

This sort of scenario probably describes this situation as well. The brother happened to get a chromosome with the 0.5% of African on it and the sister got a chromosome without it. But another sibling might end up with 0.25% African.

A Little Neanderthal in Many People's DNA

Advice: If you are interested in getting a glimpse at your Neanderthalness, go for a 23andMe test.

Back in 2010 scientists managed to get a good look at the DNA from some very old Neanderthal bones. Such a good look that they were able to see that even though Neanderthals are extinct, a bit of their DNA lives on in modern humans.

Well, most modern humans. Turns out that because of how interbreeding happened between humans and Neanderthals, people from Africa do not usually have any Neanderthal DNA.

You can figure out how Neanderthal you are with a 23andMe test. Here are what my results look like:

So it tells me how many places in my DNA I share with Neanderthals and how that compares to my relatives and other 23andMe customers. Not particularly useful but very fun!

They also display my data another way:

Markers tested for Neanderthal ancestry	Markers where you have two Neanderthal variants	Markers where you have one Neanderthal variant	Your Neanderthal Variant Total
1436	12	224	248

These are my 22 pairs of chromosomes along with my X and my Y. The dark gray marks are my Neanderthal DNA!

Now as I said, if I were of African descent, odds are there would be none of those darker marks. My DNA would be Neanderthal-free.

But it would not be human-ancestor-free. There was undoubtedly interbreeding in Africa between humans and their distant relatives. Just not with Neanderthals.

Here is why.

Modern humans started out in Africa. About 100,000 years or so ago, groups of these humans began leaving for the Middle East.

These groups then continued to spread across Europe, Asia and Australia and eventually to the Americas. But they weren't spreading into places without human relatives, just to places without modern humans.

Turns out that the ancestors of Neanderthals, *Homo erectus*, left Africa much earlier in a series of migrations. The one that led to Neanderthals is thought to have happened around 600,000 or so years ago.

The exact date isn't as important as the fact that it happened a long time before what were close to modern humans left Africa.

In that time Neanderthal DNA and the DNA of the folks left behind in Africa diverged. They built up different sets of DNA mutations until they were different enough that we can tell them apart by looking at their DNA.

So what happened then is that around 100,000 years ago humans left Africa and reintroduced themselves to their long lost relatives, the Neanderthals. The humans bred with them long enough for Neanderthal DNA to take root in human DNA.

After all this time, there are only tiny bits scattered throughout modern DNA. But this is more than people from Africa have.

The people who stayed behind in Africa didn't have the opportunity to breed with Neanderthals. That is why they do not have Neanderthal DNA (unless they have people from outside of Africa in their family tree).

And it isn't just Neanderthal DNA either. We can also see another ancient relative's DNA, Denisovan DNA, in many people (with Pacific Islanders having a lot of it).

No test yet for this long lost relative's DNA but keep your eye out. It will probably only be a matter of time until one will be available.

Tracing Your Ancestry Way Back in Time

Advice: Only get mitochondrial DNA (mtDNA) and Y-DNA tested for very specific purposes. They are not all that useful for recent ancestry or for finding long lost relatives.

The ancestry tests we've talked about up until now can usually only go a few hundred years into the past. A second kind of test can go back much further. It can go back thousands or even tens of thousands of years.

Since this sounds way more powerful than the ancestry testing I've talked about so far, you may be wondering why people don't just use these tests. They don't because it only lets you see a sliver of your entire family tree.

You can look so far back because mitochondrial DNA (mtDNA) and DNA from the Y chromosome (Y-DNA) don't recombine. Since there is none of this DNA swapping, the mtDNA you got from your mom is pretty much the same as hers. And the same as your mom's mom. And so on.

The same is true for the Y that gets passed from father to son. It is pretty much the same generation after generation because it doesn't recombine.

This means that mtDNA and Y-DNA stay very similar for long periods of time. Which is why we can trace these two DNAs much further back in time.

Of course if neither really changed at all, everyone would have the same mtDNA and Y-DNA. Since we do not, that must mean they can change. And they do.

But the changes don't happen because of recombination. Instead, the changes go by a much slower process—mutation.

Let's use our monks and their books analogy to see why not having recombination means we can trace things back in time so far.

The analogy starts out the same as when I brought it up before but with a twist. Or perhaps a tear/paste.

In this analogy we think of your DNA like a really long book.

Centuries ago, before printing presses were invented, books were copied by hand by monks in monasteries. And the monks sometimes made mistakes during copying.

Imagine that a monk leaves to found a new monastery and takes one of these books with the mistake. Now every book from that monastery will have that same mistake.

The same thing happens with DNA. Mistakes can and do happen in the DNA as it is being copied inside your body. Those "mistakes," called mutations, are handed down to future generations.

If a monk copying a book in Europe makes a mistake, all of the European copies of that book from that point on will have the same mistake. But these mistakes will probably be different than the mistakes made by monks in Africa, for example.

If an archaeologist found one of these books, he would be able to trace it back to the monastery it came from. If it has the mistake that the African monk made, it came from Africa. And if it has the European mistake, it came from Europe.

Up until now this sounds an awful lot like what I went over before. The big difference between mtDNA and Y-DNA, and the rest of your DNA, is recombination. As you'll see, it means we can't trace most of our DNA as far back in time as we can with mtDNA and the Y chromosome.

For the rest of our DNA, something closer to reality would be if the monk always had two copies of the book, maybe one with the mistake and one without it. And before he passes it on to the next monastery, he takes pages from each to create a new book.

For example, he might take the first half of the first book and combine it with the second half of the second one. Or maybe just chapter 1 from the first book and the rest from the second book. Or basically any other possible combination.

Ok so now he has a book that may or may not have the mistake. Because of this page shuffling, some of the time you couldn't tell that the book came from the European monastery even though parts of it did.

This is what happens with 22 of people's pairs of chromosomes and the pair of X's in females. This DNA swapping makes it hard to go too far back in time.

Since this doesn't happen with mtDNA and Y-DNA, we can trace it much further back. But because of the way each is inherited, neither can tell you the complete story.

Remember, fathers pass their Y to their sons and moms pass their mtDNA to their kids. This means daughters cannot see their father's Y-DNA history in their DNA. And no kids can see their dad's mtDNA history.

This results in an inheritance pattern like this:

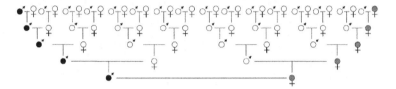

In this image, males have an arrow like this:

And females have the little plus sign like this:

This image is going back five generations, a reasonable amount for the other ancestry tests we talked about earlier.

We are tracing back the Y-DNA in black that got passed from the great, great grandfather on the top left part of the family tree. As you can see it got passed down from father to son in each generation.

We do the same thing with the mtDNA, shown in gray, from the great, great grandmother on the top right. Here I am just showing it getting passed from mother to daughter but sons get it too. But these sons do not pass on mom's mtDNA and so it is lost with them.

A quick look shows that we are missing a whole lot of our family with this test. You can only see the ancestry on the sides. The whole middle area is invisible!

So if you are male, you can see your paternal grandpa but not your maternal grandpa with Y-DNA. And you can't see your maternal grandfather either.

And the situation is similar with mtDNA. You can see your mom and your maternal grandma but not anyone on dad's side of the family nor your maternal grandpa.

It looks like all you can do is trace that one line back on each side. But it isn't as bad as this.

For example, you would also see all of your dad's brothers and mom's brothers and sisters. But you wouldn't see your mom's brother's kids. They'd have their mother's mtDNA.

In the next chapter I will explore these limits even further by focusing on a specific case. An African-American woman who hoped to find her African heritage using her mtDNA but instead found her European heritage.

What Do You Mean Not African?

An African-American woman did an African mitochondrial DNA (mtDNA) ancestry test in hopes of pinpointing a tribe of her African lineage through a company that claimed to be able to do just that. Her results said she was of European not African descent.

She was very confused because she certainly looked to be of African heritage (as did her parents and grandparents before them). She wanted to know what was going on. Was she really not African in her DNA?

After the last section you can probably guess what happened here. She undoubtedly has plenty of African in her DNA, just not in her mtDNA.

As I have said, a mtDNA test can look deep into the past which is why it is so useful for the kind of information she was looking for. But its big disadvantage is that it can only follow her maternal line back. And in fact, it can really only trace back a single maternal line.

The mtDNA is passed from mother to children. So you get your mtDNA from your mom, who got it from her mom and so on all the way back to Mitochondrial Eve (see a later chapter for more on her!).

This obviously means that the test ignores her dad's side of the family since she does not have his mtDNA. But it also means that it ignores her mom's dad's mtDNA because her mom only got hers from her mom.

And it ignores lots of other relatives from her mother's side of the family too. Pretty much anyone not on a direct maternal line will be missed.

It also means that it takes just a single ancestor from a different ethnic group to move the line onto a whole new track.

Imagine that ten or fifteen generations back, one of her ancestors along the maternal line was Caucasian. Now as we trace the line back, we are tracing her line back. She would look Caucasian.

This would be true even if everyone after that woman had kids with African American men. There wouldn't be any dilution of the mtDNA for the 200-300 years as long as there was a direct maternal link back to her.

And of course we don't need to go back ten or fifteen generations. Imagine a man like former President Obama took a mtDNA test.

His results would come back as 100% European, no African whatsoever because his mom is Caucasian. This is even though he is obviously half African from his dad.

Something like this is probably what happened in this case. There is probably an unbroken line back to a Caucasian woman and so this African-American woman's mtDNA looks Caucasian.

This kind of result is always a risk with mtDNA tests. They are incredibly powerful and incredibly limited at the same time. They can tell you a lot about distant relatives but you can only see a small subset of them.

DNA tests that look at the rest of your DNA, the autosomal DNA tests we've already covered, are also incredibly powerful and incredibly limited at the same time. While they let you look at all of your direct relatives, they can only go back a limited number of generations.

So autosomal DNA tests would definitely find African in her DNA but they wouldn't be able to see as far back in time because of how DNA is passed down.

What this means is that an autosomal DNA test would immediately have found that she was African but it may not have been able to tell her the tribe(s) her ancestors came from. And it would probably have missed her white ancestor too if she were far enough in the past!

Let's take a deeper look at all of this.

As I said, most of our DNA gets diluted over the generations. We share 50% of our DNA with our parents, around 25% with our grandparents and so on.

This is not the case with the Y chromosome or the mtDNA. The Y is passed from fathers to sons, virtually unchanged, generation after generation. And mtDNA is passed from mothers to children in the same way.

Let's look in more detail at a case where someone would be over 99.9% African but look European in a mtDNA test.

Imagine it is 1813 and a child is born of a white mother and a black father. We will draw this out like so:

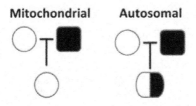

In this diagram, men are squares and women are circles. Also, European DNA is shown in white and African DNA in black.

On the left, we will follow the mtDNA. As you can see, mom passed hers on to her daughter and so both are represented with white circles.

On the right we will follow the rest of the DNA (the autosomal DNA). Here the daughter has half her DNA from mom and half from dad. I am showing this with a half white and half black circle.

Now let's say the daughter marries an African-American man and they have a daughter. Here is what that would look like:

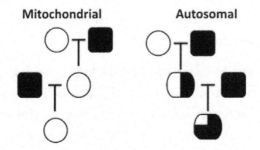

It is lined up the same way. Notice that on the mitochondrial side, she is still 100% European but is only 25% European on the autosomal side.

Let's keep going for a five or six generations (sorry my PowerPoint skills are a little rough):

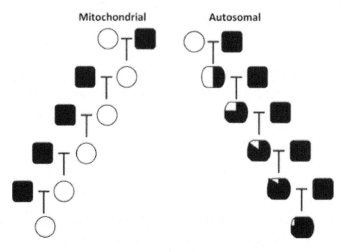

As you can see, on the left the great, great, great granddaughter still looks European with a mtDNA test. But on the right, you can barely see it anymore.

It is now 1913 and we are getting close to this woman's situation. Some people alive today will remember the woman from 1913 and all the subsequent kids and they will all be remembered as black. This is even though back in 1813, their direct maternal link was white.

Of course just because her mtDNA looks European, that does not mean that she is not African. She is as you can see on the autosomal DNA side! Heck, even the first image from back in 1813 showed half African heritage!

So buyer beware. DNA ancestry tests can be very useful if you know going in what their limitations are. But don't overinterpret them…take any results with a grain of salt.

Even if you don't find what you expected, that can be cool too. Her mtDNA test let her know about a long lost ancestor that she may not have known about at all. Perhaps a story of forbidden love from hundreds of years ago.

Identifying Richard III and Two Romanovs

Because mitochondrial DNA (mtDNA) and DNA from the Y-chromosome (Y-DNA) can be traced so far back in time, they can be very useful for seeing if you are related to anyone famous. 23andMe used to offer this but now you have to take your results and google it yourself.

These kinds of tests aren't just for frivolous uses though. They can also be used to figure out who was on a plane that crashed or if those bones you found really do belong to a king of England from hundreds of years ago.

Two famous examples are Tsar Romanov's family killed in 1918, and a king of England from the 1400's, Richard III.

Richard III. Let's start with Richard III, the last king of the York family who died in the War of the Roses. His death is often linked to the end of the Middle Ages in England and he is the subject of the Shakespeare play *Richard III*.

A set of bones was found under a parking garage that was in the right spot to belong to Richard III. One way to get more evidence that this might be him was with DNA testing.

Richard III lived so long ago that most of his DNA would not be that useful for identifying him.

Although we've seen this isn't entirely accurate, one way to think about how DNA is passed down is that it gets diluted 50% each generation. So you share 50% of your DNA with your parents, 25% with your grandparents and so on.

If we say there are five generations each century, then anyone alive today is 30 generations removed from Richard III. Using the 50% number, that means even direct descendants of Richard III would only share about 0.0000001% of Richard III's DNA.

There is no way we'd see that! And luckily we don't have to.

We can turn to those two small bits of DNA I introduced earlier—the mtDNA that is passed from mother to child and the Y-DNA that is passed from father to son. Each of these DNAs passes to the next generation virtually unchanged.

So all we have to do is follow those DNAs back in time from someone alive today back to Richard III. Easier said than done.

If you've read the previous sections, you already know that not every relative of Richard III will share these two specialized bits of DNA.

To follow the Y chromosome, you need an unbroken male line. And to follow the mtDNA, you need an unbroken female line. It takes a lot of detective work and a bit of luck to find such a relative.

For example, if we are following the mtDNA, then we need to start out with one of Richard's sisters or his mother.

He would not pass his mtDNA down to any of his direct descendants. Let's say we start out with his sister.

Next we would need to find his sister's daughters (Richard III's nieces). If a sister had only sons, then that whole line is a dead end in terms of mtDNA.

But if we find a niece, we next need to find a niece that had daughters. And so on for thirty or so generations.

If all these lines eventually end in only males, then we may need to go back up the family tree past Richard III. Maybe we need to focus on sisters of Richard III's mother and follow those lines. Or sisters of Richard III's mother's mother (his maternal grandmother). And so on again.

The researchers managed to find a man in Canada who is predicted to share mtDNA with Richard III based on his genealogy and family trees. When his DNA was compared with that of the skeleton's, the DNA matched. (This was also true of a second woman who shared mtDNA but wanted to remain anonymous.)

So the mtDNA matches. But this alone would not be enough to positively identify the body as lots of people in the past would share this DNA.

No, they needed more evidence than this to say that the body was that of Richard III. And they found it.

As I said, the skeleton was found in the general area where Richard III was reported to have been buried. They also used carbon dating to show the skeleton was from a man who was alive sometime between 1450 and 1500. This is when Richard III died. Added to this is the fact that the skeleton had a series of wounds consistent with how Richard III was killed in battle.

All of this combined with the DNA evidence was enough to conclude that the skeleton was Richard III. A 600-year mystery has been cleared up with good archaeology and DNA science.

The Romanov family. At the end of the Bolshevik revolution in Russia in 1917, Tsar Nicholas II gave up his crown. He, his wife, their five children, and four servants were then exiled to Yekaterinburg Russia.

The entire group was said to have been executed in 1918 and buried in an unmarked grave somewhere nearby. In the late 1970's, a grave was found that contained nine bodies.

DNA testing showed that five of the bodies were related to one another and were almost certainly the Tsar, the Tsarina, and three of their daughters. (The other four were most likely the four servants who followed the Romanovs into exile.)

The whereabouts of the missing son and daughter remained a mystery. There were claims that they had somehow escaped. There were even women who claimed to be Anastasia, one of the missing children.

Then, in 2007, two bodies were found near the first mass grave. They were shown to be a young boy and girl. And subsequent DNA testing showed they were almost certainly the missing Romanovs.

Back in the early 1990's researchers did extensive DNA studies on the first five bodies. The researchers looked at three different kinds of DNA to show that:

1) The bodies were four females and one male
2) They were all related
3) The male was related to the Romanov family
4) The females were related to the Tsarina's family

Once the new bodies were found, researchers did DNA testing on them (as well as additional studies on the previous five). The researchers looked at the DNA of the two new bodies and showed:

1) They were a male and a female
2) The boy was related to the Romanov family
3) The female was related to the Tsarina's family

All of this taken together provides very strong evidence that the occupants of the two graves were the Tsar, the Tsarina, and their five children. Mystery solved.

The researchers looked at three kinds of DNA evidence to establish all of these facts. They looked at mitochondrial DNA (mtDNA), the Y chromosome (Y-DNA) and autosomal DNA. Each kind of DNA has its own advantages and disadvantages.

Each cell has a lot of mtDNA and it passes unchanged from mother to children. Y-DNA passes unchanged from father to sons. And autosomal DNA has the most information about someone's parents.

Because mtDNA passes unchanged from mother to children, it can be used to quickly identify who is related to whom in one line on mom's side of the family. In this case, six of the seven people shared mtDNA with each other. The adult male did not.

This is what is expected from a family. The mom and her children will share the same mtDNA. Dad will have his mom's mtDNA.

The mtDNA can also be used to find more distantly related people. For example, the Duke of Fife and Princess Xenia Cheremeteff Sfiri were related to Tsar Nicholas II through his mother. The adult male in the grave and these two royals share the exact same mtDNA. This definitely links the male to Tsar Nicholas II.

The children and the adult woman all share mtDNA with HRH Prince Phillip. The Tsarina was related to Prince Phillip through her mother.

So the female was of the Tsarina's family and the male was of the Tsar's family. Of course, this does not prove that this is the Tsar and Tsarina. Because mtDNA doesn't change much generation to generation, a lot of people can share the same mtDNA.

The researchers searched a number of mtDNA databases and could find no matches for the mtDNA from the grave. This suggests that the mtDNA of the Tsar and Tsarina's families is not all that common.

Even if their mtDNA was fairly common, the circumstantial evidence would still be very strong for the occupants of the graves being the Tsar and his family. If not, then the father would have to be related to the Tsar and the mother to Tsarina. And they would have had to have had four daughters and one son. And to have been killed together execution style in the right place at the right time.

The researchers also compared the Y chromosomes of the man from the first grave to the boy in the second. They matched perfectly.

The Y chromosome is like mtDNA in that it passes from generation to generation virtually unchanged. But unlike mtDNA, the Y chromosome passes from father to sons.

The fact that the man and the boy share the same Y chromosome tells us that they are both related by a common male ancestor. The simplest possibility is that they are father and son. But they technically could be uncle and nephew or even more distantly related.

The researchers compared this Y chromosome to a known distant cousin of Tsar Nicholas II. The two matched. So the man and the boy were related to Tsar Nicholas II through a paternal line.

Then they looked at autosomal DNA which just means all of the DNA that isn't mtDNA, Y or X chromosome. In other words, it is the other 22 pairs of chromosomes.

These are the chromosomes that can change from generation to generation. But they can still be useful to study if scientists just look at small parts of them.

The researchers essentially used this DNA to do a paternity test on the family. The results were that the odds of the man and the woman being the parents and the five children being theirs was around 4.36 trillion to 1. This strongly indicates that the seven bodies were all members of the same family.

So to summarize the evidence:

Two separate graves were found within 75 yards of each other. The graves contained a total of eleven bodies. DNA and anatomical evidence indicated that there were four female children, one male child, four men and two women. This was the composition of the Tsar's family and servants who were exiled together.

DNA evidence showed that seven of these people were related -- the man, the woman, the four girls and the one boy. Again, this was the composition of the Tsar's family that went into exile.

DNA evidence also showed that the man was related to the family of Tsar Nicholas II on both the mother's and the father's side of the family. And that the woman was related to the Tsarina on the mother's side of the family.

This all shows just how powerful DNA testing can be in identifying bodies. Of course there are limitations in that it can sometimes be impossible to get enough DNA to get a good read.

This isn't true of just historical mysteries either. This kind of testing can help identify victims of accidents or war too.

Mitochondrial Eve and Y-Adam

As I said, we can trace mitochondrial DNA (mtDNA) and DNA from the Y-chromosome (Y-DNA) many thousands of years back. In fact, we can trace each back to someone who lived over 100,000 years ago.

Now this was not Biblical Adam and Biblical Eve. There were other women around when Mitochondrial Eve was alive and the same is true for Y-Adam. There were plenty of men around when he was alive.

Their existence is just a natural consequence of how this DNA is passed on. And because of random chance.

Remember this is what a family tree looks like for mtDNA (gray circles on the right) and Y-DNA (black circles on the left):

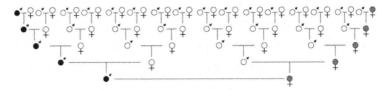

So Mitochondrial Eve is simply the last woman we can trace our maternal lines back to. She is the final gray circle as we go back in time.

(As an aside, it is important to note that this woman's mtDNA is not exactly the same as any woman's mtDNA today. We can simply trace today's DNA back to her.)

She is our mother's mother's, etc. all the way back a few hundred thousand years.

All the other women alive at the time eventually had their maternal lines end (moving forward in time). Either there was a generation with all sons or the daughters had no surviving daughters or whatever.

The descendants of most of these women still walk among us; they are just invisible to the type of testing we can do to look back so far in time. Again, this all goes for men and Y-Adam too.

What this also means is that as we look at more and more people's DNA, the identity of Mitochondrial Eve and Y-Adam will probably shift back in time. Maternal and paternal lines that were missed will be included and then we can trace all of our lines back further and further in time. In fact, this recently happened with Y-Adam.

A man in South Carolina was found to have a paternal line so different from everyone else's that he may represent a previously undiscovered branch back into our past. His DNA may change Y-Adam from a man who lived between 60,000 and 142,000 years ago to a "man" from 237,000 and 581,000 years ago.

As I have repeatedly harped on, we can't trace most of our DNA very far back at all. That old favorite recombination jumbles up our DNA between each generation.

The exception of course is the Y chromosome and mtDNA. This is why we can trace these back thousands of generations.

The Y chromosome is passed from father to son. We use this DNA to identify Y-Adam.

We use mtDNA to identify Mitochondrial Eve. This DNA is passed from mother to child (both sons and daughters).

Again what all of this means is that the existence of a Mitochondrial Eve or a Y-Adam is inevitable because of how each of these DNAs is passed down. Let's go over a couple of examples to see why.

Imagine we have our first Y-Adam, the one from 60,000-142,000 years ago. What we will do is trace his Y chromosome through a couple of generations. As I do this, I will show the X and the Y chromosomes. Remember, there are another 22 other pairs being passed down too.

Imagine he has three sons and two daughters like this:

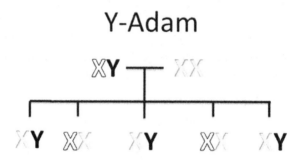

Notice that I have made the mother of Y-Adam's children have gray X chromosomes (she has two because she is a woman) while Adam has a white X and a black Y. I have done this to make tracking the chromosomes simpler.

What we can see first off is that for two lines, Y-Adam's Y has been lost. His two daughters did not get his Y chromosome and so we can't use their lines to trace things back to him using Y-DNA.

If all three sons died at this point, we could not trace Y-Adam any more (assuming he had no brothers and his dad had no brothers or that his dad's brother's had no sons or...).

Imagine the next generation went like this:

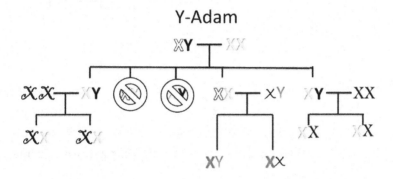

Notice that no one in the next generation has a black Y chromosome. One of his sons died before he could pass on the Y and his other two sons had daughters. (You can also see recombination happening in Adam's daughter's children's X chromosome.)

This paternal line has died out but some of Y-Adam's DNA is still in all of his grandchildren (we can't see it in his son's daughters because we aren't including the other 22 pairs of chromosomes). If we looked back using just the Y chromosome though, we would never see him.

Again this assumes that Y-Adam had no brothers. And that his father had no brothers with sons. Or that his grandfather had no brothers with sons. And so on.

Now let's look at another possible scenario:

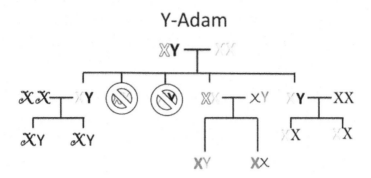

Here the paternal line lives on through the sons on the far left. He had two sons who each share their Y chromosome with their grandfather. If these sons had sons and some of those sons had sons and this kept happening, then this line could go on forever.

This is what happened in Y-Adam's case. There was at least one son to pass on that black Y chromosome all the way up to the present day, generation after generation.

Every other living male alive at the time of Y-Adam had something somewhere along the line happen where all the male lines died out. This is why we can trace the black Y all the way back to this time but not anyone else's. And not any further back either.

This exact same thing happens with mtDNA except that moms pass it to all of their children. All the women alive at the time of Mitochondrial Eve had their maternal lines die out with only Mitochondrial Eve's remaining.

Now as I said, the identity of Y-Adam and Mitochondrial Eve are not set in stone. As we look at more and more people's DNA, we may find Y chromosomes and mtDNA from branches we missed. We are not tracing back to a time when there were only two people on Earth.

If new DNA from an unexplored branch and is added to the mix, then we might push things back farther as happened with that man from South Carolina. Here is how that might look:

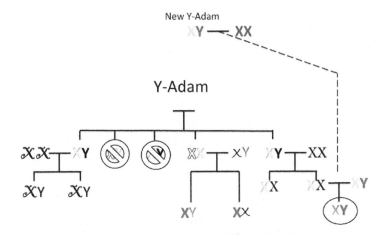

The circled man is someone who has DNA that has never been seen before. Now we need to go farther up the family tree to get to the next Y-Adam.

Bottom line is that when Y-Adam was running around, there were lots of other men there too. Same thing with Mitochondrial Eve who had lots of women sharing the Earth with her.

If we went back to their time and traced maternal and paternal lines back, we'd reach a different Mitochondrial Eve and Y-Adam. It's all relative.

Summary

See, genetics isn't as hard as your teachers made it out to be. And neither are those genetic tests.

The key complicating factor that I probably repeated too many times is recombination, that DNA swapping that happens between each generation.

It is why we can't trace most of our DNA too far back in time, why different relatives can share different amounts of DNA, a big part of why you and your sibling might have different ancestry results, and lots more.

Hopefully you also saw that ancestry results are still very much a work in progress. I focused on the basic biology but there are other factors that make ancestry results more fluid and less set in stone.

Your results are dependent on reference DNA—the groups that the company defines as Western European, Japanese and so on. As more people are added to these groups, what makes a Western European or someone with Japanese ancestors will become more refined and may change.

Ancestry results are also dependent on the algorithms that analyze your DNA. As scientists get better at writing and running these programs, your results can and will change.

In fact, there is enough wiggle room in these that if you analyze identical twin DNA (which is identical), you can get somewhat different ancestry results. The algorithms come to slightly different conclusions based on essentially the same DNA.

So take that ancestry DNA with a grain of salt. If you are 23% East European, don't conclude you are exactly 23% East European.

Instead, conclude that you have a good chunk or some or whichever qualitative word you prefer of East European ancestry. And realize that East Europe is probably a larger geographic area than you think, at least when it comes to your DNA.

Another place where genetic tests are pretty amazing these days is in relationship testing.

With the older, paternity-based tests that are still available, it is pretty hard to tell if two people are brothers let alone first cousins. The newer tests have no trouble telling a half-brother from a full-brother, a first cousin from a father and so on.

And they are incredibly powerful at telling if two people are 3rd, 4th, or even 5th cousins.

But because of the basic biology we discussed in the first section, you can't always figure out an exact relationship just from DNA. A first cousin can, for example, sometimes be hard to tell from a nephew or a grandchild.

Which is why you often need some old-fashioned genealogy work to nail down the relationship. Birth certificates and other records can help firm up a relationship.

APPENDIX A: Blood Types

Blood Type Genetics: ABO

As promised, here is how blood typing works and what blood types a child is most likely to have given his or her parents' blood type.

Let's start off with a table that shows all the possible blood types parents might have and what blood types the children might be expected to have:

Parent 1	Parent 2	Baby's possible blood types
A	A	A, O
A	B	A, B, AB, O
A	AB	A, B, AB
A	O	A, O
B	B	B, O
B	AB	A, B, AB
B	O	B, O
AB	AB	A, B, AB
AB	O	A, B
O	O	O

As you can see, two B parents, for example, are most likely to have either a B or an O child. But according to this chart, an AB or an A child is impossible.

To understand why, we need to remember three things about genetics:

1. We have two copies of most of our genes (each of those chromosomes we talked about earlier has a copy of a gene)
2. Our genes can come in different versions (called alleles)
3. Genes are recipes for proteins

There is one gene that determines the ABO part of blood type. This gene comes in three versions -- A, B, and O. Each of our two copies of this gene can be different versions.

So someone can have an A and a B, a B and an O, two A's, etc. Here are the six ways these three gene versions can be combined and what each person's blood type would be:

Genes	Blood type
AA	A
AO	A
BB	B
BO	B
AB	AB
OO	O

Some of this is pretty straightforward and makes sense without any genetics knowledge. Two of the same copies of the gene gives that blood type (two O's make O blood type for example) and an A and a B combine to make AB.

The two results that are not quite so simple are AO making A blood type and BO making B blood type. Why not AO or BO like AB?

The reason has to do with what a blood test is actually looking for. It does not look at the DNA.

Instead, it looks for the effects of you having the A or the B protein. It usually does not look at the effects of the O protein (as you'll see why below, it is actually called the H protein)

Knowing this makes the blood types easier to understand. Someone who has an A and a B version of the blood type gene makes both A and B proteins. Since the blood test looks for the effects of having these proteins, the result comes out as AB blood type.

But a blood test only looks at the effects of the B protein on someone who has a B and an O version. The result is someone with B blood type. But keep in mind that each child of this person has a 50% chance of getting O.

So what this means is that two B parents can make an O child if both parents are BO. How? By each of them passing down their O version.

If mom passes her O and so does dad, then the child will be OO which is O type blood. Each parent has a 50% chance of passing down the O gene. So each child has a 25% chance of ending up with an O blood type.

A quick way to figure this out is using one of those Punnett squares from school. The way a Punnett square works is one parent's two gene versions go on top and the other parents goes down the side. Here is an example with two BO parents:

	B	O
B		
O		

To figure out the possibilities and the chances for each, you just match up the squares like this:

	B	O
B	BB	BO
O	BO	OO

So each child has a 75% of being B and a 25% chance of being O. Note that genetically, not all of the B blood type kids are the same. The BB child will only pass down the B version of the gene. The BO child can pass on their O gene version.

As I said, the blood tests used to determine blood type look at the proteins made and not the DNA. What this means is we can't tell a BO from a BB with these tests. You could with a DNA test but these aren't routinely given.

But if this was all there was to the biology of blood type, then we could rule out a man as a dad in certain cases. For example if mom is O and dad is B, they could not have an A child. And if one parent is AB and one is O, they could only have A or B kids. None of the kids would have the same blood type as mom or dad.

This is why these tests can often be used in paternity cases. If a potential dad is AB and the child is O, odds are this man is not the dad.

And yet in rare cases, that AB parent and that O parent can have an O or an AB child. There are a couple of ways this can happen.

The cis-AB version (allele)

One way this can happen is with something called the cis-AB allele.

Remember when I said there were three versions of the ABO gene—A, B, and O? Well, it turns out there is a very rare fourth version with that awful cis-AB name. This version looks like both A and B in a blood test.

The reason this version isn't often mentioned is that with the exception of certain people with an Asian background, it is exceedingly rare. And it isn't even particularly common there.

For example, one estimate I saw stated that about 0.03% or 3 out of every 10,000 Koreans has this blood type. And that is the group where it is most common!

The next most common group is the Japanese. There it looks like about 0.001% of folks have the cis-AB allele. Or 1 out of every 100,000.

So even though it is more common for an AB parent to have an O child amongst the Koreans and Japanese, it still isn't that common. And it is much, much less common in other ethnic groups.

Still it can and does happen. Even though your high school biology teacher said it was impossible...

Remember, an AB parent can't usually have an O child because they have an A and a B allele. The parent does not have an O to pass on to the next generation and a child needs to get an O from both parents to be O.

Here is what the Punnett square looks like for most cases with one AB parent and one O parent:

	A	B
O	AO (A)	BO (B)
O	AO (A)	BO (B)

I have drawn this out so you can see both the genes, without parentheses, and the resulting blood type, in parentheses. As you can see, these two parents should only have an A or a B child.

With cis-AB, the A and the B versions are stuck together. In other words, the AB travels all at once.

Here is one example of a Punnett square of someone with cis-AB:

	[AB]	O
O	[AB]O (AB)	OO (O)
O	[AB]O (AB)	OO (O)

Here these two parents can only have what were thought to be "impossible" kids. They can only have AB or O kids.

Being cis-AB doesn't mean you always have to have impossible kids. The second copy of the ABO gene could be A, B, or O giving different possibilities.

But what is clear is that sometimes, although very rarely, an AB parent can have an O child because of the rare cis-AB allele. And this isn't the only way to get an impossible child.

Bombay blood group

The Bombay blood group is a rare blood type that was first identified in India (back when Mumbai was called Bombay). People with this blood type look O no matter what their ABO gene says.

The reason for this has to do with how A and B proteins actually get made.

You see, the A version of the ABO gene actually tweaks a protein that is already made, the H protein, into the A protein. The B version tweaks the H protein to the B version and the O version leaves the H protein alone.

What happens with the Bombay blood group is that the H protein never gets made. This means that there is nothing for the A version or the B version to work on so A, B and O all look the same.

And since a conventional blood test calls a lack of A and B as O, someone with Bombay blood group looks O even if they have an A and/or B version of the ABO gene.

So a Bombay parent will look O but could have any of the six possible genetic combinations for the ABO gene. Let's imagine a Bombay parent with an A version of the ABO gene and a B version of the gene.

He looks O but is actually AB. In the Punnett square I show this with an asterisk next to the A and B.

If the second parent were O, for example, here is what the Punnett square would look like:

	A*	B*
O	AO (A)	BO (B)
O	AO (A)	BO (B)

Remember, with a blood test this looks like two O parents (the A and the B are hidden) having A and/or B kids. These two "O" parents can only have "impossible kids" (when it comes to blood type).

Other possibilities

These are just two possibilities...there are many others. All rare but definitely all possible.

But I've probably gone on long enough about blood type (as you can see, I love genetics a bit too much!). Or have I?

I have included a section that goes a bit deeper into the science behind Bombay blood group. Feel free to skip unless you are really interested.

Deeper Science Dive into Blood Type

People with the Bombay blood group have no working copies of a gene called FUT1. And as you might have guessed, FUT1 is responsible for getting the H protein made.

So as I said before, the A version and the B version actually tweak the H protein which is made by the FUT1 gene. So the A version turns the H protein into an A and the B version turns it into a B. It looks something like this:

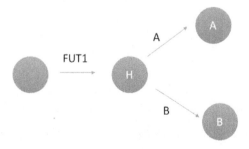

Here I have the FUT1 gene making the H protein (the circle with the "H"). Then the A allele converts the H protein to the A protein and the B allele converts the H protein to the B protein.

With Bombay blood group, FUT1 doesn't work anymore:

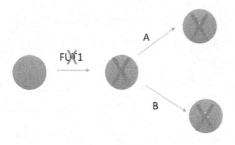

Now no H protein gets made which mean there is nothing for the A and B genes to work on. So no A or B protein gets made.

Appendix B: Dangers of Converting Raw DNA Data to Health Data

Advice: Be very careful when converting your raw results into health results. Your results may vary depending on which ancestry DNA testing company you used. And the health data you get back can be very difficult to interpret—it is hard to know what is important and what isn't.

DNA ancestry tests like we've been talking about are a lot of fun and in some cases can be incredibly useful. For example, if you're adopted, you can find out things about your past you had no way of finding out before.

As we talked about earlier in the book, you can also find long lost relatives or confirm relationships you weren't sure about. And of course, it is just plain fun to find out your family history.

But this isn't enough for everyone. After finding a seemingly endless supply of fifth cousins, you may now want to get more out of your DNA test.

You may want to learn a bit about what your DNA can say about your current and future health and maybe even about your earwax and why your pee smells funny whenever you eat asparagus.

While 23andMe offers some of these for an added cost, there is another way to find out—an online site called Promethease. But before paying the money ($10 at the writing of this book), you need to think about some of the possible issues associated with using this program.

As I said, Promethease can turn raw DNA data from companies like 23andMe, FamilyTreeDNA, and/or Ancestry.com into DNA health data.

One nice feature is that after 45 days, all the online traces of your report disappear a la Snapchat. In other words, there is no record of your health data floating around the internet for bad folks to somehow use against you. (You can download a hard copy to your computer and should within the 45 days.)

Sounds like a perfect way to convert your ancestry data into health data! Except, of course, that you need to be careful about the data you get and what you do about it.

While I haven't gone into this much in this book, it is important to keep in mind that genetic tests in general can only tell you what scientists know about the DNA these tests look at. They obviously can't tell you anything about the DNA the test doesn't cover nor things hidden in the DNA that scientists haven't figured out yet. Your results will be constrained both by what we don't yet know and by the DNA the companies happen to test.

Now none of this says anything about Promethease at all. They provide the most up-to-date information based on the literature that is available for various DNA differences that result in increased disease risk, eye color prediction, and so on.

It is just that the amount of data can be overwhelming and it can be hard to tell which bits are really important and which ones you can ignore. And the picture you will get will be incomplete and different based on which company's test results you use.

This may all sound pretty abstract but this stuff matters. To show you how, I'll use my Alzheimer's risk as an example.

My original results from 23andMe are different from the Promethease results I get from the same 23andMe data. And the results from my Ancestry.com data are different from both of these.

I got my 23andMe test before the FDA halted them from giving out this kind of health data. (The FDA has given them permission again at the writing of this book but they had not yet released their new reports.)

As you can see below, my risk for Alzheimer's, the upper risk, is much lower than average:

2.6 out of 100 men of European ethnicity who share Douglas Starr's genotype will develop Alzheimer's Disease between the ages of 50 and 79.

Average
7.2 out of 100 men of European ethnicity will develop Alzheimer's Disease between the ages of 50 and 79.

This decreased risk is almost completely determined by my having two copies of the APOE2 version of the APOE gene.

There are many other genes that will affect this number either positively or negatively. But APOE is the most important gene scientists have identified so far.

When I send my 23andMe data through Promethease, I do not get my results distilled this way. I see a much longer list that includes many findings that 23andMe probably correctly chose not to include because the results are too new, not significant enough, not done with enough participants or any other number of potential problems. This makes the report harder to interpret as I don't know which results are the most important (although the report does give me some idea).

What this means is that I see lots of reports based on different parts of my DNA. A quick look suggests a wash—the different bits of DNA I have that contribute to my Alzheimer's risk seem to all cancel out.

With more digging I would be able to glean the fact that my APOE2 is the key part of my test results and that I have a lowered risk based on the DNA they tested and by what we know about Alzheimer's and genes right now. But it would take a lot of time and not necessarily be easy (especially if I weren't already a scientist).

My Alzheimer's risk is very different when I look at my Ancestry.com results because they happen not to include the APOE markers. Because of this, I end up looking like I am at a higher risk for Alzheimer's because of all of those other, less significant markers.

If I were just to use my Ancestry.com results, I would come to a completely different conclusion about my chances for getting Alzheimer's. This result would matter a whole lot more if instead of two copies of APOE2, I had two copies of APOE4. Then I would be at a significantly higher risk for Alzheimer's but this would be invisible in my Ancestry.com results.

None of this says anything bad about Ancestry.com. They presumably chose the DNA they wanted to focus on based on what would give them the best results for ancestry and there is no reason to think they chose poorly. It is just they did not happen to choose the specific bit of DNA that indicated I had a lowered risk for Alzheimer's.

This all points to the bigger problem with trying to predict disease risk for complicated diseases with an incomplete understanding of our genome. Even if we sequenced every last A, T, C, and G, we still might not get an accurate read on our risks for having a heart attack or ending up with diabetes.

Image and links

Links:

Page 6: Random card generator:
https://www.random.org/playing-cards/

Page 21: ISOGG Wiki
http://isogg.org/wiki/Autosomal_DNA_statistics

Pages 39, 49: AAAB Accredited sites
http://www.aabb.org/sa/facilities/Pages/RTestAccrFac.aspx

Images:

Cover: DNA Helix
Tech Museum of Innovation: https://www.thetech.org/

Page 12: Chromosome image:
https://upload.wikimedia.org/wikipedia/commons/5/53/NHGRI_human_male_karyotype.png

Page 24: Cell image:
https://upload.wikimedia.org/wikipedia/commons/thumb/4/40/Simple_diagram_of_animal_cell_(en).svg/2000px-Simple_diagram_of_animal_cell_(en).svg.png

Page 76: IBD Table
Source: AncestryDNA

Made in the USA
Monee, IL
19 May 2021